An Outsider's Guide to Humans

An Outsider's Guide to Humans

What Science Taught Me About What We Do and Who We Are

CAMILLA PANG, PhD

VIKING

VIKING

An imprint of Penguin Random House LLC
penguinrandomhouse.com

First published as *Explaining Humans* in hardcover in Great Britain by Viking,
an imprint of Penguin General, a division of Penguin Random House Ltd., London, in 2020.

First American edition published by Viking in 2020.

LIBRARY OF CONGRESS CATALOGING-IN-PUBLICATION DATA

Names: Pang, Camilla, author.
Title: An outsider's guide to humans : what science taught me about what we do and who we are /
Camilla Pang, PhD.
Description: New York : Viking, 2020. | Includes index.
Identifiers: LCCN 2020027445 (print) | LCCN 2020027446 (ebook) |
ISBN 9781984881632 (hardcover) | ISBN 9781984881649 (ebook)
Subjects: LCSH: Social norms. | Human behavior. | Manners and customs.
Classification: LCC HM676 .P37 2020 (print) | LCC HM676 (ebook) | DDC 306—dc23
LC record available at https://lccn.loc.gov/2020027445
LC ebook record available at https://lccn.loc.gov/2020027446

Printed in Canada
10 9 8 7 6 5 4 3 2 1

To my mother Sonia, father Peter and sister Lydia

Contents

Introduction

It was five years into my life on Earth that I started to think I'd landed in the wrong place. I must have missed the stop.

I felt like a stranger within my own species: someone who understood the words but couldn't speak the language; who shared an appearance with fellow humans but none of the essential characteristics.

In our garden at home I would sit in a multicoloured tent tilted sideways – my spaceship – with an atlas laid out in front of me, wondering what it would take to blast off back to my home planet.

And when that didn't work, I turned to one of the few people who maybe did understand me.

'Mum, is there an instruction manual for humans?'

She looked at me blankly.

'You know . . . a guidebook, something that explains why people behave the way they do?'

I can't be certain – picking up on facial expressions was not, is not and never has been my forte – but in that moment I think I saw my mother's heart break.

'No, Millie.'

It didn't make sense. There were books on almost everything else in the universe, but none that could tell me how to *be*; none that could prepare me for the world; none that could teach me to place a comforting arm around the shoulder of one in distress, to laugh when others laughed, to cry when others cried.

I knew I must have been put on this planet for a reason and, as the years passed and my awareness of my conditions developed and my interest in science grew, I realized it was this. I would write the manual that I had always needed – one that explained humans to others like me who didn't understand, and which would help those who thought they understood to see things differently. The outsider's guide to life. This book.

It didn't always seem obvious, or achievable. I'm someone who was reading Dr Seuss while revising for my A levels. Reading fiction actually makes me afraid. But what I lack in almost everything else, I make up for with the distinctiveness of how my brain works, and my overwhelming love for science.

Let me explain. The reason I never felt normal is because I'm not. I have ASD (autism spectrum disorder), ADHD (attention deficit hyperactivity disorder) and GAD (generalized anxiety disorder). Together, these might combine to make life as a human impossible. It's often felt that way. Having autism can be like playing a computer game without the console, cooking a meal without pans or utensils, or playing music without the notes.

People with ASD have a harder time processing and understanding events on an everyday scale: often we have no filter in what we see or say, get easily overwhelmed, and can display idiosyncratic behaviours that mean our talents can be overlooked and ignored. I'm someone who will tap the table in front of me a lot, make weird squeaking noises, and twitch constantly, nervous tics assailing me at random. I'll say the wrong things at the wrong time, laugh at the sad bits of films and ask constant questions through the important parts. And I'm never far away from a total meltdown. To get an impression of how my mind works, think of the Wimbledon tennis final. The ball – my mental state – is being rallied back and forth, faster and faster. It's bouncing up and down, side to side, constantly in motion. Then, all of a sudden, there's a change. A player slips, makes an error or outwits their opponent. The ball spins out of control. A meltdown begins.

Living like this is frustrating, but also completely liberating. Being out of place also means you are in your own world – one where you are free to make the rules. What's more, over time I have come to realize that my curious cocktail of neurodiversity is also a blessing, one that has been my superpower in life – equipping me with the mental tools for fast, efficient and thorough analysis of problems. ASD means I see the world differently, and without preconceptions; while anxiety and ADHD allow me to process information at rapid speed, as I pogo between boredom and intense concentration, and mentally envisage every possible outcome of each situation I find myself in. My neurodiversity created so many questions about what it meant to be human, but it also gave me the power to answer them.

I've sought those answers through the one thing in life that gives me the greatest joy: science. Where humans are ambiguous, often contradictory and hard to understand, science is trustworthy and clear. It doesn't lie to you, mask its meaning or talk behind your back. At the age of seven, I fell in love with my uncle's science books, a source of direct, concrete information I simply couldn't find elsewhere. Every Sunday I would go up to his study and immerse myself in them. It was like a pressure valve being released – for the first time in my life I had found something to help explain the thing that confused me most: other humans. As someone who has constantly sought certainty in a world that often refuses to provide it, science has been my staunchest ally and most trusted friend.

And it's provided the lens through which I now see the world, explaining many of the most mysterious aspects of human behaviour that I have encountered during my adventure on Planet Human. While science may seem abstruse and technical to many, it can also illuminate the most important things in our lives. Cancer cells can teach us more about effective collaboration than any team-building exercise; the proteins in our bodies offer a new perspective on human relationships and interaction; and machine learning can help us to make more organized decisions. Thermodynamics explains the struggle to create order in our lives; game theory provides a path through the maze of social etiquette; and evolution demonstrates why we have such strong differences in opinion. By understanding scientific principles, we can better understand our lives as they really are: the source of our fears, the basis of our relationships, the functioning of our memory, the cause of our disagreements, the instability of our feelings and the extent of our independence.

Science has been the key to unlocking a world whose door was otherwise closed to me. And I believe the lessons it has to teach are important for all of us, whether neurotypical or neurodivergent. If we want to understand people better, then we actually need to know how people work: the functioning of the body and the natural world. The biology and physical chemistry that most of us have only glimpsed as diagrams in a textbook actually contain personalities, hierarchies and communications structures all of their own – reflecting those we experience in everyday life, and helping to explain them. Trying to understand one without the other is like reading a book with half the pages missing. A better understanding of the science that underpins our humanity, and the world we live in, is essential to a clearer understanding of ourselves and those around us. Where we normally rely on instinct, guesswork and assumptions, science can bring clarity and provide answers.

I was someone who had to learn people and human behaviour as a foreign language. By doing so, I have recognized that those who claim to be fluent have gaps in their vocabulary and understanding too. I believe this book – the instruction manual I had to create for myself through necessity – can help everyone to better understand the relationships, personal dilemmas and social situations that define our lives.

Since I can remember, my life has been dominated by one question: how do you connect with other people when you're not wired to do so? I'm someone who doesn't instinctively know what love, empathy and trust feel like – but I desperately want to. So I have become my own living science experiment: testing the words, behaviours and ways of thinking that will allow me to become, if not completely human, then at least a functioning member of my own species.

In this quest I've been fortunate to have the love and support of my family, friends and teachers who looked out for me (contrasted with others, who you will read about, who did the opposite). Because of all the privileges I have had in life, I want to share my experiences of what is possible, and what can be achieved from a starting point of difference. With my Asperger's syndrome, often referred to as a high-functioning form of autism that makes you too 'normal' to be stereotypically autistic, and too weird to be neurotypically normal, I see myself as an interpreter between both worlds in which I have lived.

I also know that what changed my life was being aware that I was seen and understood. Realizing that I was a person, and had the right to be myself: in fact the duty to be. Everyone has the right to human connection – to be heard and taken seriously. Especially those who, by nature and instinct, struggle to connect. I hope through all the experiences and ideas I share in this book, I will be able both to emphasize the importance of common ground between us as people, and to offer new thoughts on how to achieve it.

So I invite you to join me, on this journey into the strange world of my Aspergic, ADHD brain. It's an odd place to be, but certainly never dull. As well as a notebook, pack your headphones – mine rarely leave my ears, a useful barrier between me and the sensory overload of the outside world. And with that, you're ready. Let's go.

1. How to (actually) think outside the box

Machine learning and decision making

'You can't code people, Millie. That's basically impossible.'

I was eleven, and arguing with my older sister. 'Then how do we all think?'

It was something I knew instinctively then, but would only come to understand properly years later: the way we think as humans is not so different from how a computer program operates. Every one of you reading this is currently processing thoughts. Just like a computer algorithm, we ingest and respond to data – instructions, information and external stimuli. We sort that data, using it to make conscious and unconscious decisions. And we categorize it for later use, like directories within a computer, stored in order of priority. The human mind is an extraordinary processing machine, one whose awesome power is the distinguishing feature of our species.

We are all carrying a supercomputer around in our heads. But despite that, we get tripped up over everyday decisions. (Who hasn't agonized over what outfit to wear, how to phrase an email or what to have for lunch that day?) We say we don't know what to think, or that we are overwhelmed by the information and choices surrounding us.

That shouldn't really be the case when we have a machine as powerful as the brain at our disposal. If we want to improve how we make decisions, we need to make better use of the organ dedicated to doing just that.

Machines may be a poor substitute for the human brain – lacking its creativity, adaptability and emotional lens – but they can teach us a lot about how to think and make decisions more effectively. By studying the science of machine learning, we can understand the different ways to process information, and fine-tune our approach to decision making.

There are many different things computers can teach us about how to make decisions, which I will explore in this chapter. But there is also a singular, counter-intuitive lesson. To be better decision makers, we don't need to be more organized, structured or focused in how we approach and interpret information. You might expect machine learning to push us in that direction, but in fact the opposite is true. As I will explain, algorithms excel by their ability to be unstructured, to thrive amid complexity and randomness and to respond effectively to changes in circumstance. By contrast, ironically, it is we humans who tend to seek conformity and straightforward patterns in our thinking, hiding away from the complex realities which machines simply approach as another part of the overall data set.

We need some of that clear-sightedness, and a greater willingness to think in more complex ways about things that can never be simple or straightforward. It's time to admit that your computer thinks outside the box more readily than you do. But there's good news too: it can also teach us how to do the same.

Machine learning: the basics

Machine learning is a concept you may have heard of in connection with another two words that get talked about a lot – artificial intelligence (AI). This often gets presented as the next big sci-fi nightmare. But it is merely a drop in the ocean of the most powerful computer known to humanity, the one that sits inside your head. The brain's capacity for conscious thought, intuition and imagination sets it apart from any computer program that has yet been engineered. An algorithm is incredibly powerful in its ability to crunch huge volumes of data and identify the trends and patterns it is programmed to find. But it is also painfully limited.

Machine learning is a branch of AI. As a concept it is simple: you feed large amounts of data into an algorithm, which can learn or detect patterns and then apply these to any new information it encounters. In theory, the more data you input, the better able your algorithm is to understand and interpret equivalent situations it is presented with in the future.

Machine learning is what allows a computer to tell the difference between a cat and a dog, study the nature of diseases or estimate how much energy a household (and indeed the entire National Grid) is going to require in a given period. Not to mention its achievements in outsmarting professional chess and Go players at their own game.

These algorithms are all around us, processing unreal amounts of data to determine everything from what film Netflix will recommend to you next, to when your bank decides you have probably been defrauded, and which emails are destined for your junk folder.

Although they pale in insignificance to the human brain, these more basic computer programs also have something to teach us about how to use our mental computers more effectively. To understand how, let's look at the two most common techniques in machine learning: supervised and unsupervised.

Supervised learning

Supervised machine learning is where you have a specific outcome in mind, and you program the algorithm to achieve it. A bit like some of your maths textbooks, in which you could look up the answer at the back of the book, and the tricky part was working out how to get there. It's supervised because, as the programmer, you know what the answers should be. Your challenge is how to get an algorithm to always reach the right answer from a wide variety of potential inputs.

How, for instance, can you ensure an algorithm in a self-driving car will always recognize the difference between red and green on a traffic light, or what a pedestrian looks like? How do you guarantee that the algorithm you use to help diagnose cancer screens can correctly identify a tumour?

This is classification, one of the main uses of supervised learning, in which you are essentially trying to get the algorithm to correctly label something, and to prove (and over time improve) its reliability for doing this in all sorts of real-world situations. Supervised machine learning produces algorithms that can function with great efficiency, and have all sorts of applications, but at heart they are nothing more than very fast sorting and labelling machines that get better the more you use them.

Unsupervised learning

By contrast, unsupervised learning doesn't start out with any notion of what the outcome should be. There is no right answer that the algorithm is instructed to pursue. Instead, it is programmed to approach the data and identify its inherent patterns. For instance, if you had particular data on a set of voters or customers, and wanted to understand their motivations, you might use unsupervised machine learning to detect and demonstrate trends that help to explain behaviour. Do people of a certain age shop at a certain time in a certain place? What unites people in this area who voted for that political party?

In my own work, which explores the cellular structure of the immune system, I use unsupervised machine learning to identify patterns in the cell populations. I'm looking for patterns but don't know what or where they are, hence the unsupervised approach.

This is clustering, in which you group together data based on common features and themes, without seeking to classify them as A, B or C in a preconceived way. It's useful when you know what broad areas you want to explore, but don't know how to get there, or even where to look within the mass of available data. It's also for situations when you want to let the data speak for itself, rather than imposing pre-set conclusions.

Making decisions: boxes and trees

When it comes to making decisions, we have a similar choice to the one just outlined. We can set an arbitrary number of possible outcomes and choose between them, approaching problems from the top down and starting with the desired answer, much like a supervised algorithm: for example, a business judging a job candidate on whether they have certain qualifications and a minimum level of experience. Or we can start from the bottom, working our way upwards through the evidence, navigating through the detail and letting the conclusions emerge organically: the unsupervised approach. Using our recruitment example, this would see an employer consider everyone on their merits, looking at all the available evidence – someone's personality, transferable skills, enthusiasm for the job, interest and commitment – rather than making a decision based on some narrow, pre-arranged criteria. This bottom-up approach is the first port of call for people on the autistic spectrum, since we thrive on bringing together precisely curated details to form conclusions – in fact we need to do that, going through all the information and options, before we can even get close to a conclusion.

I like to think of these approaches as akin to either building a box (supervised decision making) or growing a tree (unsupervised decision making).

Thinking in boxes

Boxes are the reassuring option. They corral the available evidence and alternatives into a neat shape where you can see all sides, and the choices are obvious. You can build boxes, stack them and stand on them. They are congruent, consistent and logical. This is a neat and tidy way to think: you know what your choices are.

By contrast, trees grow organically and in some cases out of control. They have many branches and hanging from those are clusters of leaves that themselves contain all sorts of hidden complexity. A tree can take us off in all sorts of directions, many of which may prove to be decisional dead ends or complete labyrinths.

So which is better? The box or the tree? The truth is that you need both, but the reality is that most people are stuck in boxes, and never even get onto the first branch of a decision tree.

That certainly used to be the case with me. I was a box thinker, through and through. Faced with so many things I didn't and couldn't understand, I clung to every last scrap of information I could get my hands on. In between the smell of burnt toast on weekdays at 10.48 a.m. and the sound of schoolgirls gossiping in cliques, I would engage within my recreational equivalent – computer gaming and reading science books.

Night after night, throughout the years of boarding school, I would revel in my solitude by reading and copying selective bits of texts from science and maths books. My trusty instruction manuals. I took great pleasure and relief from doing this over and over, with different science books, not knowing why but only to reach the crescendo of pinning down some gravitational understanding of the reality before me. My controllable logic. The things I read helped give me rules that I set in stone, from the 'right' way of eating to the 'right' way to talk to people and the 'right' way to move between classrooms. I got stuck in a rut of knowing what I liked and liking what I knew – regurgitating a series of 'should's to myself because they felt safe and reliable.

And when I wasn't sitting with my books, I was observing: memorizing number plates on car journeys, or sitting around dinner tables contemplating the shape of people's fingernails.

As an outsider at school, I would regularly use what I now understand to be classification to understand new people entering my world. Where were they going to fit into this world of unspoken social rules and behaviours that I struggled to understand? What group would they gravitate towards? Which box could I put them in? As a young child I even insisted on sleeping in a cardboard box, day and night, enjoying the feeling of being cocooned in its safe enclosure (with my mum passing biscuits to me through a 'cat flap' cut in the side).

As a box thinker I wanted to know everything about the world and people around me, comforting myself that the more data I accumulated, the better decisions I would be able to make. But because I had no effective mechanism for processing this information, it simply piled up in more and more boxes of useless stuff: like the junk that hoarders can't bear to throw out. I would become almost immobilized by this process, at times struggling to get out of bed because I was so focused on what exact angle I should hold my body at. The more boxes of irrelevant information piled up in my mind,

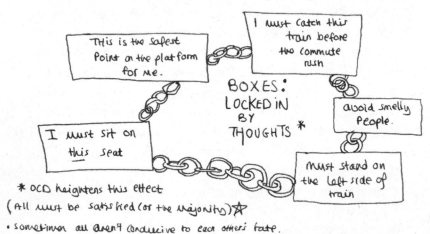

the more directionless and exhausted I became, as every box in my mind started to look the same.

My mind would also interpret information and instructions in a wholly literal way. One time I was helping my mum in the kitchen, and she asked me to go out and buy some ingredients. 'Can you get five apples, and if they have eggs get a dozen.' You can imagine her exasperation when I returned with twelve apples (the shop had indeed stocked eggs). As a box thinker, I was incapable of escaping the wholly literal bounds of an instruction like that, something I still struggle with today: such as my belief, until recently, that one could actually enrol at the University of Life.

Classification is a powerful tool, and useful for making immediate decisions about things, such as which outfit to wear or what film to watch, but it places severe limitations on our ability to process and interpret information, and make more complex decisions by using evidence from the past to inform our future.

By trying to classify our lives, thinking in boxes, we close off too many avenues and limit the range of possible outcomes. We know only one route to work, how to cook just a few meals, the same handful of places to go. Box thinking limits our horizons to the things we already know, and the 'data' in life we have already collected. It doesn't leave much space for looking at things differently, unshackling ourselves from preconceptions, or trying something new and unfamiliar. It's the mental equivalent of doing exactly the same thing at the gym every session: over time your body adapts and you see less impressive results from your workout. To hit goals, you have to keep challenging yourself and get out of the boxes that close in on you the longer you stay in them.

Box thinking also encourages us to think of every decision we make as definitively right or wrong, and to label them accordingly, as an algorithm would tell the difference between a hamster and a rat. It leaves no room for nuance, grey areas or things we haven't yet considered or found out: things we might actually enjoy, or be good at. As box thinkers, we tend to classify ourselves in terms of what we like, what we want in life and the things we are good at. The more we embrace this classification, the less willing we are to explore beyond its boundaries and test ourselves.

It is also fundamentally unscientific, letting the conclusions direct the available data, when the opposite should be true. Unless you truly believe you know the answer to every question in life before you have reviewed the evidence, then box thinking is going to limit your ability to make good decisions. It can feel good to have clearly delineated choices, but that is probably a false comfort.

That is why we need to think outside the boxes we mostly use for decision making and learn a thing or two from the unsupervised algorithm (or, if you like, go back to our childhood and climb some trees).

You might be surprised that I am recommending a messy and unstructured method over a seemingly neat and logical one. Wouldn't a scientific mind be naturally drawn towards the latter? Well, no. In fact, the opposite. Because while a tree might be sprawling, by that nature it is a far truer representation of our lives than the sharp corners of a box. Although box thinking was comforting to my ASD need to process and hoard information there and then, over time I have come to realize that clustering is by far the more useful way of understanding the world around me and navigating my way through it.

We are all wading through inconsistency, unpredictability and randomness – the things that make life real. In this context, the choices we have to make aren't often binary, and the evidence we have to consider doesn't stack up in neat piles. The clear-cut edges of the box are a reassuring illusion, because nothing is that straightforward. Boxes are static and inflexible, where our lives are dynamic and constantly changing. By contrast trees keep on evolving, just like we do. And their many branches, compared to the box's few edges, allow us to envisage many more different outcomes – reflecting the multiplicity of choice we all have.

Crucially, the tree is ideally equipped to support our decision making because it is scalable. As a fractal, which looks the same from a distance as it does up close, it can serve its purpose however large and complex the question. Like clouds, pine cones, or that Romanesco broccoli we all look at in the supermarket but never buy, it preserves the same structure regardless of scale and perspective. Unlike the box, which is limited by its form to an entirely fleeting relevance, the tree can branch out from place to place, memory to memory and decision to decision. It functions across different contexts and points in time. You can be zooming in on a single issue or trying to plot the course of your entire life. The tree will still retain its essential shape, and remain your trusted ally in decision making.

Science teaches us to embrace complex realities, not to try and smooth over them in the hope that they go away. We can only understand – and then decide – if we explore, question and reconcile things that don't fit neatly together. If we want to be more scientific about how we make decisions, that means embracing disorder before we can detect patterns and hope to draw conclusions. Which means we need to think more like trees. Let me show you what that looks like.

Thinking like a tree

Tree thinking has been my salvation. It is what allows me to function in everyday life, doing what might seem normal tasks to most of you – like commuting to work – but which could easily be insurmountable barriers for me. I can be sent into meltdown by anything from an unexpected crowd, noise or smell, to something that doesn't turn out as I had planned.

But while my ASD means I crave certainty, it doesn't mean that simplistic methods of making decisions are helpful for me. I want to know what is going to happen, but that doesn't mean I am prepared to accept the most straightforward route from A to B (and from experience and perpetual anxiety, I know the route is never that easy). It's the opposite, because I struggle to stop my mind racing through all sorts of possibilities based on everything I see and hear around me. In my world, appointments get missed, messages are left unanswered and my sense of time disappears because I've spotted something like a blackbird sitting on a roof, and wondered how it got there, and where it's going next. Or I've become distracted because I noticed the pavement was smelling like raisins after a rain shower, only to then experience a close shave with a lamp post.

What you notice is only the half of it. My mind is a kaleidoscope of future possibilities about what I observe and experience. This is why I have a whole bunch of coffee-shop loyalty cards, all fully stamped, but never yet used. I can't decide which is the greater risk: that there will be a time in the future when I need them more than I do now; or that the chain in question will cease to exist before I get the chance to use them. The net effect is that nothing at all happens. (But note: I don't consider any of these far-out projections as wrong. They are things that haven't yet happened, but still might.)

Add to that my ADHD, which means my perception of time is squished and stretched, and can sometimes disappear completely. Because information is flying through your mind at high speed, leaving your legs restless and shaking, it can feel like you are living a week's worth of thoughts and emotions all within an hour: oscillating wildly from euphoria to despondency, thinking things are going to be brilliant one moment, and a catastrophe the next. Not ideal for making to-do lists.

For the same reason, I rely on a chaotic working environment to be productive. I will spread paper everywhere, make notes on anything that comes to hand, and simply let the material pile up around me, embedded within the white noises of the room. This 'chaos' is something I find stimulating, a weed whacker to cut through the non-stop noise within my mind, enabling me to focus. In contrast to what we are taught at school, I find silence doesn't help me focus, but instead creates a pressure that simply stops me from doing anything.

My brain is craving certainty and feeding on chaos all at the same time. To keep myself functioning, I have had to develop a technique that satisfies both my need to think through everything, and my desire for an ordered life in which I know exactly where and when I am going to be. Which is where the trees come in.

A decision tree allows me to reach a certain end – which might be one of several potential outcomes, but at least I know what they are – through sometimes chaotic means. It provides structure to what I know my mind will do anyway, which is race through endless possibilities. But it does this in a way that leads me to something useful: a conclusion about what decisions I can make that give me something representing certainty. It also allows me to avoid putting all of my eggs in one basket, a process which at times gives me an outwardly cool edge of slight indifference.

Think about your morning commute. Mine is across London by train. That, for me, is an anxiety attack waiting to happen. The crowded carriage, the noise, the smells, the pressured spaces. A decision tree helps to minimize the potential for all those things to trigger a meltdown. I know what train I am going to catch, and then I consider what I'll do if it's late or cancelled, or I get delayed. I know where I want to sit, and what I will do if those seats are occupied, or if it is too loud. I think through all the things I need to ensure a meltdown-free journey – the right time, before the commuter crush; the right seat, away from the smelliest parts of the train; the right place on the platform to stand – and then I shimmy up the branches that stem off each, for when any of those things might become impossible. I am a puppet on strings of probability, which guide me forward like a harness, allowing me to manoeuvre between branches. Rather than having a fixed routine, something brittle that will break under stress, I have multiple decision trees for my commute. I have lived in my mind all sorts of scenarios, most of which will never come to pass, in the hope that I won't encounter one that hasn't occurred to me, and which is likely to freak me out.

Before I get to any decisions, the ones that reassure me it's safe to travel, I have to go through this messy mental planning. The apparent chaos of the decision tree is necessary to

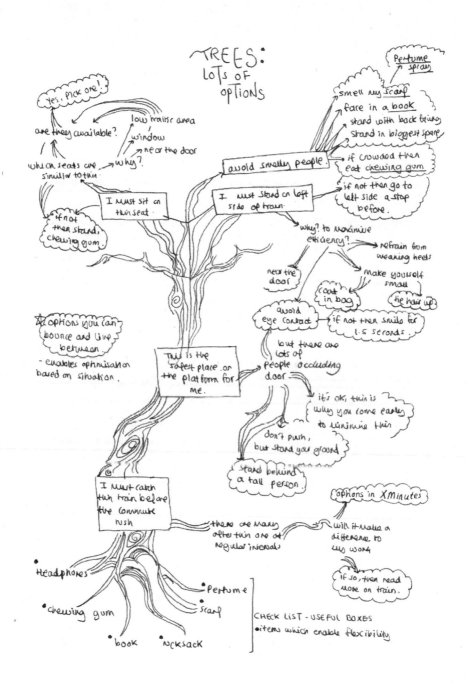

TREES:
LoTs of
OPTIONS

Perfume spray

yes, pick one!

are they available?

low traffic area
window
near the door
why?

which seats are similar to this

I must sit on this seat.

if not then stand, chewing gum

avoid smelly people.

smell my scarf
face in a book
stand with back facing
stand in biggest space

if crowded then eat chewing gum

if not then go to left side a stop before.

I must stand on left side of train.

why? to maximise efficiency?

refrain from wearing heels

near the door

Make yourself small

coat in bag

tie hair up

avoid eye contact

if not then smile for 1.5 seconds.

* options you can bounce and live between
- enables optimisation based on situation.

This is the safest place on the platform for me.

but there are lots of people occluding door

it's ok, this is why you come early to minimise this

don't push, but stand your ground

stand behind a tall person

options in X minutes

I must catch this train before the commute rush

there are many other train are at regular intervals

will it make a difference to my work

if so, then read more on train.

• Headphones

• chewing gum

• book

• Perfume

• Scarf

• rucksack

CHECK LIST - USEFUL BOXES
• items which enable flexibility

help get me towards the sense of certainty I need to function.

To you that probably sounds like a lot of hassle (you'd be right!) and, to be clear, I'm not suggesting that you start war-gaming your morning routine as I would. I need to do this, otherwise I would get too overwhelmed and just not leave the house. But I do think this method has a place when it comes to the more complex decisions – the ones that neuro-typical instinct and methods tend to fall down on.

While the challenge for my ASD/ADHD brain is not get-ting paralysed by overthinking, the opposite of this is also a problem. If you don't delve deeply enough into the data set that surrounds every major decision, allowing yourself to consider the different possibilities and outcomes, and the branches of the tree that different decisions will simultan-eously close off and open up, then you are effectively making your choice while blindfolded. We can't predict the future, of course, but for most situations we can cluster enough data points and plot enough possibilities to give ourselves a decent map. What I do to reassure myself and tamp down anxiety on an everyday basis could be useful for you in working through the difficult decisions in your life. With a decision tree, you can reach from the things that you know, to grasp onto the decisions you are seeking – not in a prescriptive way, based on fixed outcomes, but by letting the evidence guide your con-clusion, and allowing yourself to consider multiple outcomes and their implications.

Trees are also necessary to make sense of the confusing, open-ended questions that people are so fond of asking. If someone asks me, 'What do you feel like doing today?' my instinctive response will be, 'I don't know, maybe.' I need some specific options – branches of a tree – to offer a route from the chaos of total freedom towards the restriction of a decision, one which still leaves open alternative routes to divert down. A tree turns the multitude of underlying events and variables inherent in any decision into something like a route map. It might make every conversation into a jungle trek, but at least it allows me to find a way through.

By contrast, when we make decisions based on box thinking, we are usually doing so through a combination of emotion or gut instinct. Neither can be relied upon, and you can take it from me: there's nothing like ADHD to help you understand what making immediate decisions after emotion has slapped you in the face is like. Good times.

Good decisions don't generally emerge from an assumption of certainty, but out of the chaos otherwise known as evidence. You need to start from the bottom, building upwards towards the conclusion rather than starting with it. And to do that, you need a tree to climb up.

So how do I decide?

A tree is all very well in theory, you might be thinking, but with so many branches how do I actually make a decision? Isn't there a risk of getting lost in all the wonderful complexity we have visualized around us?

Yes, there is (welcome to my world!) but don't worry, machine learning has your back again. Algorithms also have a lot to teach us about how to sift through large amounts of data and draw conclusions – exactly what you need to do to make the tree method work for you in everyday circumstances.

Any machine-learning process essentially begins with what we call feature selection: filtering the useful data away from the noise. We need to narrow down our evidence base and focus on the information that can lead us somewhere. This is about setting the parameters of the experiments you will then undertake.

How is this done? There are different methods, but one of the most common in unsupervised machine learning is known as 'k-means clustering'. This is where you create indicative clusters within a data set based on how closely related they are. Essentially, you group together things that look similar, or have certain features in common, to create a certain number of clusters, and then use those to test and evolve your assumptions. Because you don't know what the outcome is meant to be, you are open-minded about the conclusions, and initially focus only on what can be inferred from the data: letting it tell its own story.

Is that really so different from the decisions we have to make all the time? Whether they are insignificant or life-changing choices, we always have data points that we can examine and try and cluster together. If it's choosing an outfit, we know what makes us feel good, what is appropriate for the occasion, and what others might think. If it's deciding whether to take a job in a different country, the data points might range from the salary on offer, to lifestyle, proximity to friends and family, and career ambitions.

If, when facing a big decision, you've ever said to yourself, 'I don't know where to start', then feature selection is not a bad place – even if the choice is daunting, it enables you to consider many alternative possibilities, leaving you in a stronger and more empowered place.

First you separate the things that really matter from those that are just distractions – the main determinant being how it will make you feel, now or in the future. Then you group together those that share common features: what helps get me from A to B, or fulfils a particular need or ambition. And with those clusters, you can start to build the branches of your decision tree and see how your data points are related. This process helps to reveal the real choices you face, as opposed to those which are in the forefront of your mind to begin with (perhaps driven by FOMO – fear of missing out – or the likely judgement of strangers on social media). Those factors exist on their own tree separate from you, and simply cannot be compared like with like.

It's never really a choice about the red top or the black, or about this job or that one. Those are just symbols and expressions of the things you actually want. Only by sorting the data and building your decision tree can you see how to navigate the choices in front of you, and reach decisions based on meaningful outcomes; for example, will it make me happy and fulfilled?

It's always more complex than the binary 'yes or no' decisions that we like to pretend exist. We need to go deeper than the immediate choice point and mine the data of our emotions, ambitions, hopes and fears about the decision ahead, understanding how they are all linked, and what leads us to where. By doing this we can be more realistic about what a particular choice will and won't achieve for us; deciding important things based on the fundamentals of what matters most in our lives, and less aligned with the boxes strewn around us. They simply represent our emotional baggage and immediate instincts, often populated in the piled boxes of social 'should's ('I *should* explore the world while I'm young,' 'I *should* settle down and not take that risky job abroad,' and so on) on how to be and behave. Variances in mental health are often perceived as a losing battle in this regard, since they naturally push and challenge such boxes.

We can also learn from the machine-learning process about how evidence should be used. Feature selection and k-means clustering get you to the starting line, but that's all they do. Reaching conclusions requires a whole extra phasing of testing, iterating and refining. Evidence in science is something to be tested, not waved around like the tablets of stone. You make assumptions so you can question and improve them, not treat them as an immutable guide for life, however concrete they may seem.

We should treat the evidence in our own lives on the same basis. By all means pursue one branch of the tree that seems most favourable, but don't saw off all the others before you do (the essence of declaring one option 'right' and the other 'wrong'). Experiment with what you think you want, and be willing to backtrack and adjust your assumptions if it doesn't work out as expected. The beauty of a tree-like structure is the ability to move easily between branches, whereas traversing between seemingly unrelated boxes leaves us anxious and without a clear path forward, resulting in an inevitable retreat. Just as any data set contains a mixture of inherent patterns, hidden truths and total red herrings, our lives comprise a whole selection of pathways forward, forks in the road and cul-de-sacs. Sorting the evidence can give you a good idea of which to pursue, but don't bet against having to double-back and try again. Life isn't linear but branched, and we need our thought patterns to match that reality.

That may sound haphazard, but it's actually a far more scientific and sustainable method than making a decision and sticking to it regardless of the evidence. It allows us to set the course of our lives in the same way as machines are engineered: with more precision, and greater willingness to test, learn and adjust. It's also something that we improve at over time: as we get older, we gather ever more data, allowing us to grow more mature, complex trees in our heads that better reflect the reality of a situation – like an architect's drawing of a house compared to a child's.

The good news is, you are probably doing a bit of this already. Social media has made scientists of us all when it comes to the art of posting the perfect photo. What angle, what combination of people and objects, what time of day and what hashtags? We observe, test, try again and over time perfect a method for documenting our perfect lives for the world to see. And if you can do it on Instagram, you can do it in the rest of your life too.

Learning to embrace error

By taking this approach to decision making, building chaos and complexity into our mental model through a tree-thinking or unsupervised approach, we start to develop a more realistic method for predicting events and making decisions based on the evidence available.

This method isn't just useful because it's scalable, flexible and more clearly represents the complex reality of our lives. It also equips us better to respond when things go wrong – or in situations where we think they might have done.

This is the point, to put it bluntly, at which the scientific approach copes a lot better than people. When you are a biochemist or a statistician, error doesn't faze you, because you can't afford to let it. It can be exasperating and time-consuming, but it's also essential and fascinating. Science thrives on error, because it allows us to fine-tune, to evolve and to fix mistakes in our underlying assumptions. Only through the anomalies and the outliers can we reach a full understanding of the cell, data set or maths problem we are studying.

It's why statistics uses standard error as a basic principle, building in an assumption that there will always be things that don't accord with expectations and predictions. In machine learning, we have 'noisy data', information that sits in the data set but doesn't actually tell us anything useful or help inform the creation of meaningful clusters. Only by acknowledging the natural noise in the system can we facilitate performance in big data groups. You can't optimize unless you study and understand the noise, errors and deviations from the mean. After all, noise in one context is often a signal in the next – much as one person's trash is another's treasure, since signals are not objective, but based on what an individual is looking for. If scientists didn't embrace the need for error, and find fascination in things that contradict and frustrate their assumptions, ground-breaking research would simply never happen.

People, on the other hand, can be less sanguine when things don't go according to plan. You won't find many commuters cheerfully quoting standard error when their train gets delayed or cancelled. This is because we have been taught to consider mistakes through an emotional, not a scientific, lens. We are generally quick to declare errors as symptomatic of category failure, and conclude that the system doesn't work, or the decision that led to this point was entirely wrong. The truth is usually more mundane: the trains do run on time, most of the time; and your decision might have worked differently in the vast majority of foreseeable scenarios.

Experiencing a setback of some sort is not sufficient evidence to conclude that everything has failed, or that a system or decision should be abandoned wholesale. Humanity would have achieved only a fraction of what it has if that approach to error had been used by scientists and technologists throughout history. Even in everyday life, it is when things go wrong that people come alive – albeit that might be an angry reaction to a train being late, or when a stranger has stolen your trusted waiting spot.

The knee-jerk response to error is one of the main downfalls of box thinking. By acting much like supervised algorithms, we assign a binary quality to every data point and situation. Yes or no. Right or wrong. Rat or hamster. This limits our ability to see problems in their proper context, and makes every error seem like a critical one. The train is cancelled therefore my day is ruined. It creates the dangerous illusion that there is always a categorically right or wrong decision, and that the difficult thing is making a cliff edge decision between them (pure box thinking). In my case, it also tends to lead to one setback – the missed train – derailing my entire day, sending me cascading into meltdown because my plan has gone awry.

Because reality is more nuanced than that, so must our techniques be for thinking about problems and reaching decisions. With a box, you have nowhere to go when something goes wrong. Your only choice is to label it as a failure, and start again. With a tree, you are surrounded by alternative branches: paths forward that you have gamed out in your head. It's far easier and more efficient to switch course because you didn't put all your chips on one outcome working out as you had hoped. You have already planned for just this eventuality, and left yourself with plenty of worthwhile backups.

Counter-intuitively, machine learning can help us to be less mechanical, and more human, in how we assess the decisions in front of us. It teaches us that 'mistakes' are normal and are inherent in real data. There are few, if any, truly binary choices, and not everything fits into a pattern or can be tied up into a neat and irrefutable conclusion. The exception makes the rule. I've benefited from the machine-learning perspective not because it filters out the randomness and uncertainty that are an inherent part of humanity, but because it embraces them more easily than most people do, and provides a method for assimilating them. It allows me to plan for situations I know I will find intimidating, and be better prepared for when things go wrong.

Tree-like thinking is important because it reflects the complexity that surrounds us, but also because it helps us to be resilient. Like a mighty oak that has stood for hundreds of years, a decision tree can stand up to all weathers, long after a box has been jumped on, broken and cast aside for ever.

2.   How to embrace your weird

Biochemistry, friendship and the power of difference

To say I never fitted in at school would be something of an understatement.

Maybe it was the fact that I had a dedicated adult mentor sitting next to me in each class, my tendency to go into meltdown when a teacher said a word that scared me, or my uncontrollable nervous tics. I can't imagine my penchant for giant tubes of antiseptic cream did me any favours either.

In so many ways I was set apart from my classmates. How many schoolchildren have had to fire someone, as I did my new helper at the age of ten, based on her excruciatingly bad breath?

Because kids like nothing better than to gang up on the outsider it was often open season on me. 'You're nuts', 'She's an alien', 'You should be in a zoo.' (The last was a personal favourite.)

That sounds awful, you might be thinking. And it some ways I suppose it was. When the snide comments and the in-jokes sank in (because it usually took me a few hours to actually understand why the comment was hostile), I would put my head under the covers and howl, ears ringing and hot blood pumping to my cheeks within the soft silence of my duvet, until my face was blotchy and my hair sticky.

But in one crucial, and brilliant, way it was amazing. Because all the things that ostracized me from the social cliques of the playground also gave me an armour no one else around me had. It took me a long time to realize it, but my difference actually held one considerable advantage. Unlike pretty much any neurotypical teenager on the planet, I was immune from peer pressure. (And believe me, I tried not to be.)

This wasn't out of high-minded principle or good judgement. I didn't define myself against the social status quo; I just failed to understand it. But my lack of interest in being a part of the crowd left me free to observe its rhythms, which I did with great care. I would sit on a bench above the playground during lunch breaks, watching the different cliques and the subcultures; from those scrapping over a game of football to the busy gaggles that were always full of screaming and laughter, and the smaller groups of two and three who hung around the edges. From my perch, I could see the whole eco-system of playground species.

What I saw confused me. There were so many contradictions, not least between individual personalities and group dynamics. Why was it that people would act very differently, depending on who they were with, or the particular circumstance? Why would I see boys gravitate towards the mean of their social clique, imitating behaviour as specific as the pitch of their voice and the amount of gel in their hair? If you've ever wondered why a friend of yours has suddenly started behaving differently when around new people, you will know how I felt: confused at how a person you think you know can suddenly start pretending to be someone else.

I had no affinity for these hidden, counter-intuitive social connections. I could see the exchange of what seemed like an almost invisible friendship currency, one asymmetrical with people's own personality types: they were changing aspects of how they looked and behaved simply to mimic others they wanted to be friends with. But I couldn't understand why this was, or what compelled people to give up something about themselves to become part of a social group. Being social animals wasn't allowing the people I observed to be themselves, but actually undermining their distinctive personalities and preferences.

I wasn't getting anywhere trying to model human behaviour just by watching people. There was too much data to get a proper handle on. But then I had a breakthrough, not in the playground or studying in the chemistry lab, but while watching football one weekend in the common room.

I wasn't following the game so much as the players. Some of them were constantly communicating with the others, shouting gregariously. Others were in their own shell, focusing just on the job they had to do. There were players constantly running all over the pitch, and others who largely stayed in their set areas of the field. This was a football team, but it was also a collection of individuals, responding dynamically to a fluctuating situation, all bringing their own different skills, personalities and perspectives to the table. It was more than twenty-two men kicking a ball around a field. It was a human behaviour experiment, and one sufficiently limited to draw useful conclusions from. Much better than anything you could cook up in a test tube.

My eyes widened with delight as I reached an epiphany, a realization that this dynamic behaviour could in fact be modelled. I stood up and practically bellowed, 'They are like proteins!' Eureka! I felt as though I had just scored the winning goal, but the others didn't look ready to mob me in shared delight. Blank and uneasy faces stared back at me. 'Just watch the game, Millie.'

For perhaps the first time, I had seen human behaviour through a lens I could understand. The unusually disciplined model of the football team had reminded me of how protein molecules work together so effectively to keep our bodies functioning.

Proteins are among the most important molecules we possess, because they are also among the most collegiate. They play distinct roles in helping the body to interpret changes, communicate them and decide on actions as a result. Our bodies work in large part because our proteins know their own role, appreciate that of their peers and act accordingly. They work as part of a team, but through the expression of entirely individual personalities and capabilities. Dynamic yet defined, individual within a team context, proteins can offer a new model for how we organize and interact as people. Like humans, proteins respond to their environment, communicate information, make decisions and then put them into action. But unlike us, proteins are actually very good at doing this: working in an instinctively collaborative way without letting personality clashes, personal problems or office politics become obstacles. And they achieve this not by trying to 'fit in' with their environment, but by aligning and making use of their various chemistries: embracing the complementarity of contrasting 'types'.

The protein model of teamwork – one that makes the most of differences rather than suppressing them – is so much more powerful than the human urge towards homogeneity in social situations: the desire to fit in. How much are we losing out on by seeking to mask our distinctive skills and personalities rather than being proud of them, and making them our differentiator?

Our individual quirks and differences don't just make us who we are as people. They can also make our friendships, our social groups and our working relationships function more effectively. We should be proud to be weird in our own different ways, not only because it feels good, but because it helps things to work better. You can take that from me, someone who has learned that my ASD, ADHD and anxiety are superpowers that give me a precious and unique perspective, not the barriers that many assume. But even better than that, as this chapter will explain, you can learn it from understanding the ways in which our proteins make us tick.

The wonder of proteins

It's hard for me to express how much love I have for proteins. They are these beautifully chaotic modules of evolution, whose interwoven network of functions brings biology to life. In the same way that some children ascribe personalities to their pets or imaginary friends, and use them to start learning about human behaviour, I saw the personality in proteins. Like people, they behave in an unpredictable and non-linear way. They are dynamic, versatile and susceptible to changing conditions and interactions with others like them. Proteins are a thing close to my heart – literally.

Just like people, there is no one protein type. There are many different kinds, performing a dizzying array of functions that keep the body moving and protect us from danger. These roles are dictated by their form and structure, in the same way that humans behave differently, do a variety of jobs and perform contrasting social functions in group settings, all largely determined by personality type and life experiences. There are the protein equivalents of introverts and extroverts, leaders and followers, goalkeepers and box-to-box midfielders.

In this way, proteins reflect and can help to explain aspects of human behaviour. But that's not all. Because they don't experience peer pressure or emotional ups and downs, they can also be understood as something of an ideal of how humans could behave – since they act based on what is most energetically favourable to them, focusing on the need at hand and ᴗ not getting distracted by emotions or self-consciousness. Proteins, indifferent to micromolecular judgement, don't need to worry about fitting in with their peers or seeking uniformity. Instead, they can capitalize on and exploit their different skills, forming teams whose success is based on the power of difference.

It was no accident that proteins have formed the basis for my understanding of human behaviour. They are the fundamental element of all biochemistry. You can't understand how cells form, mutate and interact without understanding the nature and behaviour of proteins. Which means you can't understand how the body works without knowing something about protein, the most prevalent substance in our systems after water. Among their many functions, proteins form the enzymes that help us to digest food, the antibodies that allow us to fight disease, and the molecules (haemoglobin) that transport oxygen around the body. Protein is also a vital ingredient in our skin, hair, muscles and major organs.

As you can see, there is no humanity without protein to provide the building blocks. And for me a few years ago, there was no understanding humans without starting from what I knew about proteins.

Not all of my football hunches have turned out to be good ones. Supporting Manchester United hasn't looked like such a good option since Sir Alex Ferguson retired in 2013. But I wasn't wrong about the parallels between proteins and people. That realization has proved to be as important a factor in my life as Fergie was to United's long, now sadly curtailed run of success.

The four stages of protein

As well as being intrinsic to the functioning of the human body, proteins are also surprisingly similar to people in their behaviour and evolutionary development. We can start to see this by examining the different evolutionary stages of a protein molecule, and the parallels to our development as humans.

A protein starts life in its primary structure, which looks, when seen through the microscope, a bit like a cooked piece of spaghetti, tangling around in different directions. This is flexible by design, not limited to any specific structure and capable of adopting many different roles. In our digestive systems, there is no single protein that can break down everything we put in our bodies. You need a different one for each major food group: amylase to digest starch, lipase for fats and protease for proteins (yes, there is a protein for processing proteins).

Of course, as well as being component parts, proteins also have their own building blocks: amino acids. The primary structure is set by the unique order of amino acid codes predetermined by the gene sequences in our DNA – the essential coding of our physiology. Even a few amino acid variations in the hundreds that make up a single protein molecule can make a significant difference in what it will become in the cell and its external expression to others (known as the phenotype), such as eye colour.

Like us as people, the fate of a protein is to some extent coded into it from the moment of creation. And just as we adapt and change as we grow, products of both our genetics and our upbringing, so too do proteins. A protein fold and a human mind both emerge as a delicate balance of biochemical interactions, determined by a combination of the inherent sequence and the surrounding environment: the intersection of nature and nurture. A protein's initial sequencing may determine its direction, but the actual form and function only become clear at this secondary stage. For most of them, the initial 'spaghetti' structure is too unstable to allow them to function properly. In these cases, proteins progress to their secondary state, folding in on themselves to become more stable and versatile, three-dimensional structures – much like a human learning to move independently by crawling.

The development of this secondary structure is the next stage of determining a protein's purpose. Take keratin, the fibrous protein that is a major component of everything from wool and hair (which is why, if you check the ingredients carefully, you will find alpha keratins in your shampoo and conditioner) to fingernails and birds' claws. At the secondary stage, keratin either forms an alpha-helix (twisty) structure, whose compactness and rigidity makes it one of biology's stronger creations, or a structure referred to as a beta-sheet: one that's looser, flatter and softer. This you will find in spiders' webs, birds' feathers and waterproofing the skin of many reptiles.

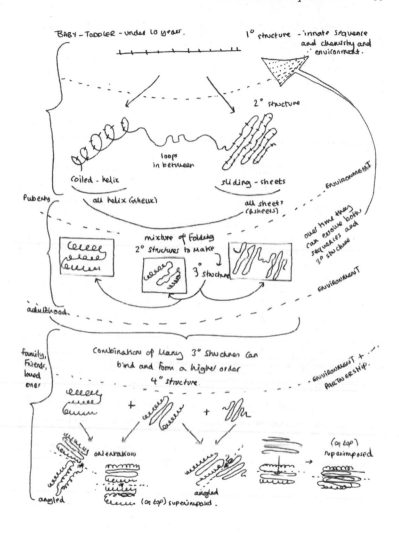

As time goes on, the secondary structure undergoes further interactions with itself, forming higher-order structures more specific to their sequence and environment. In your muscles, you have two kinds of protein: myosin (thick ones) and actin (thin ones). For a bicep to contract, these have to interact, where the myosin kicks the actin using chemical energy, causing them to slide past each other to generate the contraction. Which is why you can hold this book up with your hands to read it.

This requires further folding to produce the advanced – and specific – tertiary structure, in which proteins start to specialize, adapting themselves for a particular job just as many of us undertake professional training to become a scientist, doctor or lawyer.

The tertiary structure represents the final stage of a protein's development. By this point, it's not going to fold into anything more complex. But it does adapt, binding with different partner proteins to fulfil a variety of functions. This is what my mum refers to as people being 'cooked', coming out of the oven of personal and professional development to become fully functioning adults, ready to fly solo and take on life. For proteins and humans alike, this is the moment of self-sufficiency at which you are ready to both act independently and work effectively in conjunction with others.

This final, quaternary structure reflects not an additional stage of development, but rather all the alternative forms it can take and bonds it can form. Actin, when it's not helping your muscles to work, can also help cells adhere together and move around the body, and as such plays an important role in your immune system and in creating the cell tissue to heal wounds. It's a versatile little thing: definitely one of those hard-working midfielders on the human-body team.

Do you ever feel as though you are a different person at work and at home? That's like a protein in its quaternary structure, adapting to conditions and context, performing different roles depending on what is needed to keep the body's engine running smoothly. In this quaternary form, a protein is a model of versatility, moving from one use to the next as needed. An example to us all, and a handy way for me to understand yet another confusing aspect of human behaviour: why it isn't always consistent from one situation to the next. That said, I believe proteins do evolution of this kind much better than people: they change their form and function without reservation, where we can often get stuck in a rut and struggle to accept the need for personal growth, resisting changes in our environment rather than adapting to them as a protein would.

For me at fifteen, what had been hard to understand through watching people, was starting to become clearer by putting cells containing proteins under the microscope: observing how they evolve and grow, and noting that their interactions are dynamic and context-dependent. We scientists might like to define and categorize everything we know about proteins, and how they work, but the truth is they can be every bit as fickle, capricious and hard to pin down as the people they provide the foundations for.

That said, when it comes to group rather than individual behaviour, proteins have a major advantage. Without the distraction of emotional impulses, or worrying what others might think, they are free to organize themselves in the most objectively efficient way. A team of proteins is all action, no politics. It gets the job done. Let's look next at how that is achieved.

Protein personalities and teamwork

Most of us will recognize that our friends span a range of personality types. There are the more outgoing, the more introverted, and people who are better at communicating, taking action or showing empathy. Then there are people like me, who have had to ask how long you should hug someone to offer comfort (two to three seconds since you're asking, four if it was a really bad break-up).

We take on different roles that reflect our personalities, often without realizing it. In any given group, some people feel much more comfortable taking the lead, while others would rather someone decided for them. Some like to announce what they think directly, others just to hint at it (eugh).

None of this is by accident. From a cell organism to a workplace, there is no gathering of people, animals or molecules whose behaviour cannot be explained by some form of hierarchy and set of relationships, determined by both personality and physiology. In bee colonies, you encounter several different types of bee: the workers who build and defend the nest, while also gathering food; the queen who is the social glue and 'boss'; and the drone whose only responsibility is to mate, and is expelled by the others from the hive outside mating season. The colony therefore relies on the diversity of different bee types, their range of functions and their ability to pick up on each other's range of signals.

Like a bee colony, both cell organisms and social cliques can be understood by examining how the different components – proteins or people – communicate with each other. In the same way that a friendship group decides where to go out, or what film to watch, a cell relies on different inputs and actions, from different protein types, to perform necessary functions.

Or at least that's the theory behind efficient organization, and something we see in cell structures and the animal kingdom. Human behaviour is often a much messier reality.

Think about your own friends and how good you are at deciding how to socialize. How long does it take to agree a date, fix a venue and get everyone signed up? And how much of that process involves people doing things they don't really want to do, or at times that don't really suit them? Again, the desire for conformity and to be positively judged by others tends to override the necessity of communicating and acting effectively in concert.

By contrast, proteins are a marvel of efficient organization over emotional compromise and social politics. You can see this by looking at the process of 'cell signalling' – essentially how different proteins combine to sense changes in the body, communicate them to each other and make decisions as a result.

I used this as my model to understand what proteins could teach me about both the human behaviour I had observed, and what a better model might look like. My approach was to map protein behaviour onto the Myers–Briggs Type Indicator, which matches people's personalities to eight different attributes – Extroversion, Introversion, Sensing, Intuition, Thinking, Feeling, Judging, Perceiving – and determines the four that best reflect who we are and how we behave.

In doing so, I started to realize that proteins were an even better map for people than I had assumed. On one level, they are an effective proxy for human personality types, as the examples below will explain. But proteins don't just show the reality of how different 'types' exist together. They also provide a model for how coexistence and collaboration should really work, and the importance of expressing your personality rather than seeking to conceal it.

Here are some of the most prevalent protein personalities.

Receptor proteins

The first point of contact for any cell in the body is the receptor proteins. They sense changes in their external environment – for instance a spike in blood sugar levels – and subsequently transmit signals downstream to other protein bodies in the cell, for further processing. Think of them as the empathetic members of the group, the ones who have an instinct for when someone is feeling uncomfortable, or that an argument is about to get out of hand. They are not a decision maker, but a mediator, working with others like them.

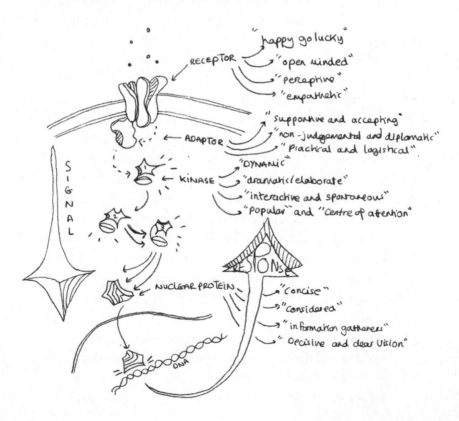

Receptors are the happy-go-lucky people who move easily between different social groups; those who can be members of multiple cliques and communicate across them. In Myers – Briggs terms they are ENFP – 'Warmly enthusiastic and imaginative, see life as full of possibilities, make connections between events and information very quickly' – or ENFJ – 'Warm, empathetic, responsive and responsible. Highly attuned to the emotions, needs and motivations of others'.

Quick to perceive, they are the diplomatic social butterflies, socially at ease with other people and good at breaking the ice.

Adaptor proteins

The adaptor facilitates the next stage of cell signalling. It binds onto the receptor, and decides the best way to communicate the message across the cell. It is a cell's first 'decision-making' body, determining which downstream, or 'kinase', protein will be activated and what message it will send to the rest of the cell. It's the adaptor which turns the initial signal into a message which can then be communicated and acted upon.

To me, these are the no-drama, chilled-out types who are good at supporting others and don't need to be the centre of attention. I often got on well with 'adaptor' type people, who are non-judgemental and very good at translating between different people and personality types in a diplomatic way. Like the receptor they are communicators, but not in the proactive, making-friends kind of way. They're more facilitators: smoothing the path towards the most desirable outcome.

They are the ESTJ – 'Practical, realistic, matter-of-fact. Decisive, quickly move to implement decisions' – or ISTP – 'Tolerant and flexible, quiet observers until a problem appears, then act quickly to find workable solutions'. The people who don't shout the loudest or push themselves to the front of the queue, but without whom a group can lose its balance and fall apart.

Kinase proteins

Once the signal hits the kinase protein (enzyme), things start to really happen. Kinases are the motivators of biochemistry. In simple terms, they catalyse the transfer of chemical energy groups to their downstream effectors and interactors, kick-starting all the necessary functions for a cell to respond to change.

'You're a bit of a kinase, aren't you?' I once said to a friend, not getting the approving reaction I had expected from a comment that was intended as complimentary and reassuring. Nor did she respond well when I followed up with a helpful explanation: 'Kinases, they are one of the most promiscuous and popular proteins in a cell.' (In technical terms, functional promiscuity is the ability of a protein to cause a beneficial side reaction in parallel to the main reaction it is catalysing. Of course, there are non-technical definitions of promiscuity too: I'll leave to your imagination how these apply to kinase-type people . . .)

Kinase people are full-on extroverts: life and soul of the party, relishing the intensity and frequency of contact with other people. They are all handshakes, hugs, pats on the shoulder and kisses on the cheek (*shiver*).

They are the social hubs and the energy givers: lovers of parties and attention. In Myers–Briggs terms, kinase people are ENTP – 'Quick, ingenious, stimulating, alert and outspoken . . .

Bored by routine' – or ESTP – 'Focus on the here-and-now, spontaneous, enjoy each moment that they can be active with others' – or ENTJ – 'Frank, decisive, assume leadership readily'.

A dominant part of social cliques, kinases were not always my cup of tea. My sensory processing couldn't hack all that expression and energy, and I would generally try to avoid them. But if you are at a party that hasn't yet sparked into action, it's probably the case that the kinases haven't yet arrived.

Nuclear proteins

Other proteins communicate and catalyse, but only nuclear proteins can act to turn a received signal into a cellular response. All the activations I've already described are leading to the proteins in the nucleus, the 'brain' of the cell that coordinates activity. It is the nucleus that defines how the cell will respond, and what is actually going to happen.

For example, if you cut yourself, and start bleeding, then the body knows it needs to repair the damaged vessels. This problem will be sensed by the receptors and passed through a series of kinases to the nuclear protein (in this case, called the HIF). It will then respond in a way that creates proteins to increase the production of blood vessels, ensuring more can flow towards the damaged cells. And that, if you'll forgive me, is bloody marvellous. The nuclear protein is the captain of the ship: knowing what buttons to press in certain situations, thanks to its iceberg watchers working away upstream. It makes sure that the information gathered by the receptors, and transmitted by the kinases, is acted upon.

Just as every cell has a nucleus, and every football team a

captain, every social clique has the individual to whom all others defer on the important decisions. These people will often be less buzzy and involved than the kinases: sitting to one side where they have a better perspective on events.

They are hyper-focused, specialist and often more intro-verted than you would expect. Myers–Briggs might classify them as INFJ – 'Develop a clear vision about how best to serve the common good. Organized and decisive in imple-menting their vision' – or INTJ – 'Have original minds and great drive for implementing their ideas and achieving their goals. They quickly see patterns in external events and develop long-range exploratory perspectives.'

A 'nuclear' person won't always, or often, be the centre of attention. But everyone acknowledges them as the boss. And their word is usually final.

As you can see, proteins are an exemplar of teamwork and effi-cient organization. Different types play distinctive roles that accord with their personalities: all of them are needed for the body to work efficiently. They don't get jealous of each other, or start coveting an alternative role. It's a low-ego, high-productivity environment. If only every workplace or friendship group could say the same.

The productive example of proteins can help us in all sorts of ways. For me, studying proteins helped shape my approach to forming relationships and handling social situations. Understanding my protein and personality types helps me to navigate people: deciphering the best way to intervene and get the outcome I want when I communicate. It means know-ing that it's easiest to strike up a conversation with a 'receptor', the ones most likely to talk to me and in the best position to pass a message on. Or acknowledging that ultimate decisions aren't usually made by the kinase who shouts the loudest, but by the nucleus who though they might look lost in their own thoughts actually wield the real authority. Human and social behaviour may look impenetrable to those of us who don't grasp it instinctively, or even those who think they do, but I have reassured myself over time that there are patterns in it to be deciphered and understood. What might sometimes look and feel random usually comes down to the different personalities in a group, the nature of their interactions and the external factors they are responding to. If you've understood your proteins, then you're a lot further on to understanding how the people around you think, act and decide.

As for me, the protein I most resemble will change according to the circumstances. By nature I am more of an adaptor or nuclear protein, observing what's going on around me more than I actively dive in. But in the right circumstances – with people I feel comfortable with, or when discussing issues I have expertise on at work – I can be as much of a fizzing kinase as the next extrovert. There's nothing that says we have to pick one style and stick to it; adaptability according to the situation is both normal and a good reflection of protein-like behaviour.

It was understanding proteins that helped me to see why other girls at school would get upset by completely different things from me – when their hair got wet in the rain, or a teacher told them to do up their top button (behaving like receptors, extra-sensitive to the outside world and its perception of them; or kinases, attention seekers). I didn't really *understand* why these were such terrible fates, but at least I was able to help prepare for them: always carrying an umbrella in case of another unexpected shower.

Proteins also helped me to realize that, when it came to 'fitting in', there was no substitute for just being myself. There was a time, in my teens, when I thought I could train myself to be like everyone else: emulating the behaviour of my peers so I could adopt their interests, their mannerisms and their language.

I wanted to infiltrate a group of girls – just for 'funsies', as they would have said – to do as they did, share the jokes they shared and get excited by the same things as them. I wanted to be basic, so badly. Of course, this began with appropriate research. Having googled 'How to be a basic bitch', I came across very specific results, describing people with an affinity for pumpkin spiced lattes, puffa jackets and small but meaningful tattoos. So I bought the jacket, I drank the coffees and I watched *Dawson's Creek* and *Made in Chelsea*, hoping somehow that these would provide the camouflage and the connection. It was when I fell asleep in front of the second of these – the big craze at the time – that I realized it wasn't working. I ended up wearing a jacket I didn't like, which limited my arm movement; drinking something I didn't want to drink; and pretending to laugh at a lot of jokes I didn't find funny (laughing in the wrong places, of course). It was exhausting: even more than ADHD itself. And most of all, I missed my science books (to this day it is being able to study maths at the weekend that makes me tick). Trying to blend in by mimicking my peers, I had ended up suppressing my own personality, an even worse feeling than that of being left out. Proteins would teach me never to try to repeat this experiment or fall victim to the siren call of conformity.

The most important lesson we can learn from proteins is about how better to interact and work with others. That's because, unlike people, proteins recognize and respect the need for difference. It's what makes them function in the coordinated way I have just described, with all the different types adopting complementary roles. People, not so much. Our group behaviour may be defined by our different personalities, but much of the human instinct is towards uniformity. Most of us are basically driven by the desire to fit in, and win acceptance from our peers. Although we fulfil different roles in social situations, this is largely done unconsciously, and we don't understand and harness these dynamics in a helpful way. What's more, our urge to belong risks being counterproductive, when it is actually our differences that define us and underpin effective communication and cooperation. Rather than denying or masking our true personality, we should embrace and harness it. If you're a listener, then make the most of being someone with high levels of empathy; or if you're one of the world's kinases, then your ability to make people laugh is probably your superpower. We would work better as humans – in both social and professional contexts – if we allowed ourselves to *be* ourselves, and were more accepting of other people doing the same.

Science shows us that, far from uniformity being helpful, it is diversity that is essential to collaboration and success. Unfortunately, one of nature's best examples of this is the cancer cell: a marvel of biological communication and inter-dependency. With a tumour, some cells work to maintain its growth, while others protect the exterior and work to neutralize the immune system and therapies.

We still have a lot to learn about cancer, but unpalatable as it may sound, we also have something to learn from it. In a tumour, there are none of the ego factors that undermine teams of people, at work or on the football field. The different cells fulfil their specific roles and can also evolve into parallel functions if the conditions demand. A tumour is a model of biological empathy: the parts subsuming themselves to the needs of the whole. This is also what makes cancer so hard to treat. There are so many different cells to target, and their ability to evolve and shift role makes them incredibly difficult to pin down. Cancer always has a next move, and researchers are constantly playing catch-up.

But if cancer can thrive – against all our best research and treatment efforts – through diversity and effective collaboration, then so can we as people, if we take seriously the need to understand the different personality types, roles and interconnections that are needed to create a functioning ecosystem. We need to understand, and embrace, our differences – our weirdness – to benefit from the kind of efficiency that biological creations enjoy by design.

Perhaps you've just started a new job and want to understand how your new office really works. Look for your proteins: separate the kinases who talk the most in meetings from the nuclei who are probably making the important decisions. Find some receptors who can help you to settle in. Identify the adaptors who don't make a fuss but are essential to getting things done.

We should be applying the same thinking to how we build teams in any field. Often companies will talk about a certain kind of person they want to recruit, as if one personality type can meet all the different requirements that a business needs to be successful. This notion of a uniform ethos to which everyone should sign up runs counter to what the cancer cell proves: that diversity, and the ability to evolve, are the most fundamental parts of achieving constant growth to stay ahead of the competition. Just as the successful football team needs players covering a whole variety of positions, a thriving organization relies on a range of characters and perspectives.

The example of proteins highlights two areas where, as humans, we often fail to fulfil our potential: evolution and the utility of difference. If we believed more in our ability to evolve and change in life – just as the protein molecule does – and trusted more in the distinctiveness of our personality and perspective (while giving equal credence to those of others around us) – we could short-circuit many of the inhibitions and misunderstandings that hold us back, both as individuals and in how we organize collectively: as friends, at home and at work.

The lesson of proteins is to be more confident, less self-conscious and more accepting of the different roles people play, because of our very different personalities. It's to curb the basic human (or at least neurotypical) impulse to fit in, and to celebrate our weirdness, recognizing its essential contribution to social cohesion. Proteins teach us that being different helps us to work well together, and that individuality is fundamental to effective teamwork. Big lessons from a molecule that you need a microscope to see. It's time more of us took a closer look.

3.   How to forget about perfection
Thermodynamics, order and disorder

'The state of your room is scary,' my mum once declared on visiting my school dorm. 'I can't even find a place to sit!'

Who hasn't argued with their mother at some point about tidying their room, and had differing interpretations of what constitutes mess?

My own cluttered kingdom was less the product of laziness than anxiety. What to the untrained eye seemed like chaos was to me personally tailored, with everything where I put it last, spontaneously curated to ensure the optimum place for immediate use. Possessions strewn across the middle of the floor hadn't just been left there by accident, but to ensure I would have equal access to them from either end of the room.

'It's spontaneous and adaptable! To me this is how I roll.' That earned me a motherly rolling of the eyes. 'Good luck with that,' she murmured, in exactly the same tone of voice she had used when I said aged four that I wanted to marry Elton John.

There was another explanation for the questionable state of my room, albeit one I didn't dare to advance in that discussion: thermodynamics. This is the science of how energy moves and is transferred. Its laws tell us that, left to itself, the universe inevitably becomes more disordered over time. All our efforts to create order are fighting the second law of thermodynamics, which dictates that the entropy (loosely, disorder) of a system will always naturally increase, as less energy becomes available to be used. So perhaps an untidy room can't ultimately be avoided even if we try.

But I don't just make this point in the hope that teenagers across the world will start quoting thermodynamic theory at their parents to justify piles of unwashed socks. An understanding of the principles might make for a good debating stance, but it's also about appreciating something more fundamental: the role that order and disorder play in our lives, and the laws of physics that govern them.

As I grappled with the state of my living space, the desire to please my mother's sense of tidiness alongside my own needs, and the question of what tidy even was let alone how it could be achieved, thermodynamic theory became my guide. It helped me to understand more closely my own desire for order, and the ways in which that can and can't be achieved: the difference between something like a weekly meal plan which just makes your evenings more efficient, and the challenge of reconciling your vision of order (for instance how to organize a room, or plan a holiday) with that of your friends, family and loved ones. And it gave me a crucial new perspective: that our efforts to create order in our lives do not exist in isolation, but in a messy context of people and inanimate objects, all with their own energetic needs.

In any friendship or relationship, you have to blend your own sense of order with that of someone else. And while this might seem like a straightforward matter of making compromises, it's often more complicated than that. Because our individual ideas about order are not straightforward, vague or lightly held: they are detailed masterpieces that have evolved from multiple layers of experience, preference and deeply ingrained habit, and which represent silent expectations that are often only voiced when they have been breached. Try to daub over that with broad brushstrokes and you will quickly run into trouble.

Unless we understand and respect these needs, recognizing the frame that thermodynamics sets for our lives, we will face an uphill struggle to achieve the equilibrium – of mood, environment and lifestyle – that we seek. We are all trying to live the life we want, the way we want to live it, at the same time as making room for other people's preferences, needs and quirks, and reconciling our own with what is realistic in the time and space available. Appreciating thermodynamics allows us to work with the grain of the world around us. It's the key to a balanced life. And a tidy room.

A disordered orderly person

For obvious reasons, it's always been important to me to have a sense of order in all aspects of my life, something that I can rely on. But unfortunately that hasn't tended to translate into a tidy living and working environment. This is quite a common and paradoxical feature of autism, especially in mine, where we crave order and certainty but often struggle to create it for ourselves.

Therefore, when we manage to find a reliable means of controlling everyday order, we hold on to it come hell or high water – be it the arrangement of food on a plate, the positioning of curtains in a room, the exact layout of a workstation, or the appropriation of a specific chair. These all form the threads of routine we attach ourselves to, and which together help to bind our everyday functioning. (Though confusingly, this is always subject to change, and often driven by passing fads of certainty that swiftly dissolve.)

But much as I love routine in all aspects of my life, I was struggling to create order within my own kingdom. To me the books and papers strewn all over, and my regal 'floordrobe', were simply a convenient, spontaneous layout that allowed me to find things quickly. At the same time, however, it bugged me that I was so overtly ordered in other parts of my life, but not in the space that was most important to me. Even though it suited one part of me to have things thrown all around, my OCD was also kicking in and demanding consistency with the ordered Millie who existed elsewhere. It felt like lying to people – something I simply cannot do – to say that I was an ordered person, when my bedroom looked like this. This inherent inconsistency unnerved me like a cricked neck, different parts straining asymmetrically against each other.

Between my own need for order, my mother's impression of what order should be and my instinct for what an efficient environment looks like, I was in conflict. And that was before I even got to the problem of how to create a 'tidy' space, and the endless permutations my brain was throwing at me for what this might look like, plus all the different routes for getting there. I had no objective idea of what 'tidy' actually looked like. As I crunched the permutations, my mind started to feel like a raisin, anxiety in full spin.

I felt blocked. This is akin to the difficulties exhibited for those on the autistic spectrum, where often the challenge isn't having no idea of what to do, but too many ideas. We envisage all the possible ways and combinations of actions of how to tidy, like a massive landscape with no filter, generally leaving us stuck, marooned in the middle. There are just too many degrees of freedom and options, combinations that pull you from all angles like puppet strings.

So yes, you might have surmised, I was finding it hard – *really* hard – to tidy my room. This remains true today, with every living and working space generally in a state of flux, even though I will freak out if someone else moves something and disturbs my fragile sense of ordered disorder. You probably know the feeling: the way you arrange your desk might not make sense to anyone else, but it works just fine for you. We always know when someone has been sitting in (and messing with) our office chair when we get back from holiday. And it's never a good day when something unexpected has messed with your routine, in however small a way. Even for people who like surprises, an underlying need for order still exists.

Even worse, my mum was due for her repeat visit in three days' time, so I knew I had to do something. It won't surprise you that, rather than seeking out any kind of guidance on home improvement, or simply throwing my floordrobe into the actual wardrobe in one go (which didn't work, by the way), I reached instead for a science textbook. *The Elements of Physical Chemistry*, to be exact (found eventually under my yoga mat – I told you the system worked).

It didn't offer any insight on how to tidy, but it does contain a more important idea that we all need to understand. 'Creating order from disorder requires energy and is not thermodynamically favourable [something that happens spontaneously without additional energy being applied, like an ice cube melting].' Tell me about it; even thinking about tidying hurts.

This was the source of my problem, and the reason we all have to work hard to create and maintain order in our lives. It's a fundamentally unnatural state, pitting your need for the neat and tidy against the universe's impulse for fast and loose. It's not random that things become less ordered over time, it's simply the destiny of molecular physics.

This is nothing less than the source of humanity's battle with nature. We build walls, only to see them crumble over time. We paint buildings, knowing that eventually it will flake away and have to be redone. We organize our possessions, realizing that they will before long become disorganized again without consistent and repeated effort.

Everything you do in your life to create order, however big or small, gets undone sooner or later. And that means you have to do it all over again to counteract disorder. As *Elements* tells us, this requires energy. The folding of clothes, the washing of dishes and shaving to name but a few. Or, in my case, trying to follow the 'rules' of everyday communication

So the next time you're struggling with how to get something the way you want it, don't blame yourself. Blame thermodynamics. And more than that, understand that thermodynamics sets the terms of how we can organize our lives. You can have order, but only through expending energy on it. And the order you create will, however carefully you plan, be reversed over time.

In the pursuit of an ordered life, we have to recognize that we are not acting in isolation. There is a whole world of molecular physics out there we have to navigate. Doing so means accepting that a degree of disorder is inevitable.

You have to pick your battles and make some compromises –
starting with your idea of the perfectly ordered life.

Why does disorder increase?

Let's start by looking a bit more closely at the laws of thermodynamics and how they ensure disorder in our lives.

There are four in total, but it's the first two that concern us
here.

The first law states that energy can neither be created nor
destroyed, it can only change its location and form.

The second tells us something about what happens to
energy as it changes form. It states that the entropy of an isolated system can only increase or stay the same. When systems
have low entropy, they have the most thermal energy available to react – so that's exactly what they do, undergoing
spontaneous reactions that are thermodynamically favourable (often from solid to liquid or gas). After this has happened,
the energy doesn't go away (remember the first law), but nor
does it remain in the same state as it was before. Once you've
burnt a log of wood and turned it into smoke and ash, the fire
goes out. Its energy is now spread in a more disparate, dispersed way, and is said to be less 'available' than it was before.
Entropy measures this availability of thermal energy to drive
reactions in systems – and the higher the entropy, the less
energy is 'available'.

What the second law tells us is that, through naturally
occurring processes, the energy in a system will always move
over time to a less ordered, less productive state – one that
allows it to do less work (aka, Tuesday evenings).

Another simple example is to think what happens to an ice cube when you take it out of the freezer. Given a little time, it melts into a liquid state before eventually evaporating into water vapour. Through both of these changes, its entropy increases. The molecules that were tightly packed together as a solid are now bouncing around, having a high old time in their gaseous state. And so entropy, and therefore disorder, has increased. This is what the second law tells us is happening all around us. (For those wondering what happens when you condense your water vapour and freeze the liquid back into ice, that is no longer an isolated system because you have applied external energy to reverse the naturally occurring change).

In very simple terms, therefore, thermodynamics tells us that entropy (disorder) is always increasing in processes that happen spontaneously, and without external intervention. This explains why it's much easier to shatter a glass into a thousand pieces (therefore increasing its entropy) than it is to put them back together again. Or, as I discovered many years ago, it takes only a second to kick a pile of leaves in a park, but you can spend the whole afternoon trying to rearrange them into their original places. For the record, you won't come close to succeeding, but you will end up after five hours absolutely knackered and in a state of delusional self-doubt. Good times.

Unlike the spontaneous reactions that increase entropy, creating order is hard work because it requires external energy. You are having to fight the impulse of a system to undergo a spontaneous reaction that increases entropy.

Measuring the energy of systems in this way (via a metric called Gibbs free energy, which calculates how much is available at a given moment) is not some arbitrary or technical matter. It's actually the most important tool we have, across almost any field of science, to understand whether something is going to happen or not. Thermodynamics governs so much of what we understand and how we research. It's an utterly reliable way of knowing whether one reaction is more likely to happen than another. We just have to ask what is more thermodynamically favourable.

I have always found thermodynamics comforting because it offers certainty, where people so often present confusion. Did someone mean exactly what they just said, or were they hinting at something else? Am I missing a nuance that is never written or spoken, but somehow meant to be understood? To me this is like a broken radio: however hard I try to tune in, the reception is always poor. By contrast, the signals thermodynamics sends us are crystal clear.

So is the lesson it offers us in everyday life. Tidying our homes isn't just difficult because it's a pain to fold and stack things, work out where everything should go and wrestle with your duvet cover. It's because you are trying to lower the entropy of an environment whose spontaneous progression is in the opposite direction, towards disorder. So when a parent, partner or flatmate tries to make you change your ways and tidy up your things, they aren't just asking you to overcome your own laziness or subvert your unique sense of order; they're also pitting you against the fundamentals of thermodynamics. A much better excuse for leaving things as they are.

Being thermodynamically favourable

As we try to fashion order in our lives, we need to understand the uphill battle that thermodynamics has created for us. But that doesn't mean we are powerless to fight it. Remember how the textbook told us that creating order from disorder is thermodynamically unfavourable? Lowering the entropy (disorder) of a system takes work and expends energy, but it's not impossible. It is just something you have to invest time and energy in achieving. So you have to decide what is worth the sacrifice. For me, this book was a good example, its ideas plucked and pruned from a tidal wave of disorder within my own head. It took a lot of energy to pin down the ideas into something ordered, but it is an exercise I have both enjoyed and learned a huge amount from doing.

You can't have your penny and the bun, so the question then becomes what is preferable: to conserve our mental and physical pennies, or to achieve the objective we have set ourselves in a given set of circumstances? The important thing is to recognize the trade-off. If we want to create order, we are going into battle to some degree against thermodynamics. It will cost us energy in some way or form.

So how do we achieve the things we want, and make the sense of order we crave, without creating too great a thermodynamic opponent to overcome?

Part of the answer is to be realistic in our expectations. The more precise your vision of order, the lower entropy state you are envisaging, and the more effort it is going to take to achieve it. Thermodynamics is the enemy of perfectionism, because the second law turns it into a Sisyphean battle.

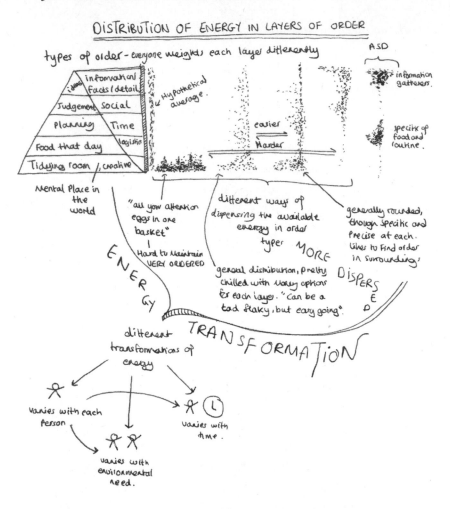

DISTRIBUTION OF ENERGY IN LAYERS OF ORDER

types of order - everyone weights each layer differently

ASD

information
Facts / detail

Hypothetical average.

Judgement   Social

Planning   Time

Food that day   logistics

Tidying room /   creative

information gatherers.

specific of food and routine.

easier
___
Harder

Mental Place in the world

"all your attention eggs in one basket"
|
Hard to maintain VERY ORDERED

different ways of dispensing the available energy in order types   MORE

general distribution, pretty chilled with many options for each layer. "Can be a tad flaky, but easy going"

generally rounded, though specific and precise at each. Likes to find order in surroundings

DISPERSED

ENERGY   TRANSFORMATION

different transformations of energy

varies with each person

varies with time.

varies with environmental need.

However close we get to rolling the boulder up to the top of the hill, the molecular drift towards disorder is always going to keep dragging it back down. The more perfect your sense of order is, the higher and sharper your hill becomes, the more thermodynamically unfavourable the situation, and the more energy it takes to even get close to the summit.

So you need to change your expectations, not so much lowering them as redistributing them. We have the same amount of energy and attention, and the question is how we utilize it. Recognize that you can't climb every pointed peak of perfectionism on the mountain, and you need to leave yourself enough energy to tackle those that really matter.

For instance, I know that tidying my room will take me two days of preparation, allowing me to work out all the permutations, deal with the anxiety that comes from having lots of options, and ask myself whether there is even such a thing as a tidy room, and what it could look like. Once the work finally gets under way, it is hardly any quicker. On the tidying mission to placate my mum, it took an hour to decide where to move my clock and mug. Two hours in and I had managed to reposition the laundry basket and open a window. My mind was spinning with all the choices and priorities: what should go where, and didn't it depend on the context? Like a computer running too many programs, my mind was becoming clogged with all the choices and options, its mental cursor lagging around the screen. Time for a cup of tea and a lie-down. Again.

As you can see, for me room tidying is about as thermodynamically unfavourable as it gets. I'm battling against not only the inevitability of disorder, but my own multi-layered perspective about what might represent order. I'm exhausted by the choices and confusion of it all, and how my invisible preferences relate to other people's.

That being the case, why do it at all? Well, because unlike thermodynamic experiments, our lives aren't isolated systems. We are living with and alongside our friends, family and loved ones, all of whom have their own parameters of what is favourable. And that means there have to be compromises. You can't just focus on what is optimal for you as an individual, you also need to understand and empathize with those around you. So there are even more battles to choose between than we first thought.

Order: competing visions

You never have to look far to find what people disagree about. If you've ever tried creating a playlist for your office, agreeing on a film that a large group of people is happy to watch, or finding an undisputed recipe for Bolognese sauce, you'll know how our sense of what is optimal varies from person to person, often significantly.

Will there ever be a point at which we all agree on whether or not tomatoes are delicious (correct answer: yes), or the fact that a tidy room is *this* configuration, or that a certain person has *these* character traits? Unfortunately, probably not.

It's no different when it comes to our visions of what order looks like. For some, colour-coding, neat piles and clean desks are preferable and even necessary, while others actively feed off a more chaotic system, or simply lack patience for all the filing. Others again, much like myself, are an inconsistent mixture of the two.

When I decided to tidy my room, even though it wasn't favourable to me, it was something I knew would reassure my mum. It was a compromise: a way of showing my love. I was satisfying her sense of order over my own. And perhaps, too, I was seeking a little of the order that she always created in our home, where everything had its clearly stated place, even down to the plug in the bathroom sink. My sink didn't even have a plug.

Whichever sense of order we choose to prioritize, the important thing is recognizing that these different perspectives exist. It's very easy to go ahead with the assumption that what works for you must also be optimum for everyone else. After all, it makes total sense to you, and it can be hard to see how it might look from another standpoint.

As someone who has relied on very specific modes of order to anchor my anxiety, I do my utmost to return this favour. I know that a lot of people supported and respected my sense of order when I was little, from having certain foods on certain plates, to an intricately planned schedule for the day, always having my hair in a tight plait (one bobble at the top and another at the bottom), and watching specific movies only after having recited my pre-show introduction, and appropriated my favourite chair.

If we want harmonious relationships, we need greater empathy with how the people around us see the world, and how their own sense of order differs from our own. It might seem like nothing if you reorganize the spice rack in a communal kit chen, move the pans to a different cupboard or switch around the arrangement of the knives and forks in a drawer. But to the people you are sharing the space with, any of those could feel like a significant change that destabilizes their sense of order, meaning they can't easily find what they're looking for. A small thing from one perspective can often seem much more serious from another. When I was moving out of my old flat and the landlord was doing viewings, I came home one day to find the blinds had been raised just a *little* too far, and that was enough in itself to trigger a mini meltdown.

If someone else attempts to impose their order on you 'for your own good' (which is actually theirs), then this is a form of controlling behaviour and a battle which you have every right not to partake in. I have definitely been guilty of this, since I had a phobia of smoking and drunk people up until I was twenty-three, and in retrospect have sabotaged friendships out of nothing more than pure fear.

On the other hand, if you love your mum enough to disturb your energetic equilibrium to make her happy, or if you manage to actively restrain a harsh response to a friend who is under stress, then I salute you. These silent, everyday sacrifices are microscopic gestures that define generosity and love between people.

That said, having empathy for the needs of others doesn't mean you should give up on your own. I know from experience the pitfalls of imitation. When I was first tussling with the problem of what a tidy room might look like, I sought inspiration by inspecting how my friends ordered their living spaces and indeed their lives. Perhaps if I mimicked their dress and eating habits, the way they arranged their wardrobes and the posters on their walls, I could tap into some sense of perfect order – literally stepping into someone else's shoes (and copying the way they put their socks on, which for the record was not consistent). This fell apart when not only did imitation fail to deliver any notable changes, but I caused consternation by copying another habit of one friend: kissing the poster of Zac Efron on her wall before going to bed. 'You don't have to fancy him just because I do,' she said, embarrassed. All I could think of was whether it was going to help me tidy my room. And then the poster was moved, and I had to start all over again.

The perils of imitation can be more serious than this. If we take our cues from other people, we will never find out what is most thermodynamically favourable for us, and how to tune into our own mental ecosystem – optimizing how we use our limited store of mental and physical energy to support our most important needs.

That might mean we go out rather than stay in because of peer pressure, real or imagined. It can affect choices in everything from the food we eat to the way we dress. All the time, we are encountering situations in which our own sense of order conflicts with that of people around us. We have to choose when to compromise, and when to stick to our own path.

I discovered the perils of following others in my late teens, when I got a bit obsessed with being healthy. What was 'good' for me and how could I live a healthy life? Searching around these questions online (always a big mistake) gave me some very clear answers. It meant doing a lot of exercise, and cutting out certain foods, instructions I followed to the letter. So much so, in fact, that I ended up eliminating almost all of the major food groups from my diet, once went three days eating noth ing except an apple, and when I turned seventeen weighed just forty kilos. I ignored the fact that I was hungry and often nauseous, because I thought I was ticking the boxes that would allow me to be as healthy as possible. It was only by pushing myself to that accidental extreme that I recognized this artificial sense of good order was actually the worst possible thing for me. But it took years to teach myself that, if I want to go to the gym several times a week, I need to eat a healthy diet that fuels my body to exercise.

My landscape is often very sharp, as someone who thinks their way up and down mountains of evidence and different options, meaning I learn from wide-ranging extremes. Like many teenagers, I found defining the shape of my preferred world hard, requiring lots of trial and error. It's not just creating order that's hard; it's the effort it takes to work out what order on your terms actually looks and feels like.

Our optimum state of being and way of living are incredibly personal. And while we need to make compromises with the people around us, understanding that their needs are as personal and deeply held as our own, we also need to keep hold of our identity – avoiding a situation where other people set the terms of how we live, and what we expend our energy on.

Equilibrium: can we achieve it?

There is one important thermodynamic concept I haven't properly introduced yet, but which can help us to navigate the order and disorder that surrounds us. Enter equilibrium.

Equilibrium is the mother of all two-sided reactions, and it also happens to be my all-time favourite. On every scale of scientific, social and psychological abstraction, there is always a form of it happening. Equilibrium explains how we can walk, breathe voluntarily and hold a book like this; it's the reason central heating warms your room and a cake can never be unbaked.

Technically speaking, it is the state reached when a chemical reaction is in balance, with both the forward and reverse sides occurring at the same rate, and the overall state of the system no longer changing. If you put a very hot object next to a cold one, equilibrium has been reached when both are at the same temperature. The seesaw of reaction has reached its perfectly balanced, lateral state.

Thermodynamic law tells us that equilibrium is the state which every isolated system is seeking to reach, because it is the most efficient, where Gibbs free energy has dwindled its efforts to zero: there is no work left to be done, and therefore no energy required.

It sounds like the ideal state of existence: everything in proportion, working away effortlessly, and with no surprises or sudden changes. The problem is that equilibrium simply isn't achievable for us as people – whether biologically or metaphorically. The closest our bodies get is through homeostasis, the series of processes that helps to regulate the body's internal environment, from temperature to water and mineral volumes and blood sugar levels. This is responsible for everything from how much we sweat, to when our blood vessels contract or dilate, and when insulin is released in the body.

But this is not total equilibrium, as I discovered in one of my uncle's science books, through a sentence that left me feeling disturbed but also liberated. It said that when the body has reached that final equilibrium with its surroundings, it is considered dead. Therefore equilibrium is ultimately the definition of human mortality. Make of that what you will, but to crave and thrive off this pursuit for something both unachievable and fatal is a curiously human condition.

In contrast to the energy-neutral state of a chemical or thermal equilibrium, homeostasis is a demanding process involving multiple organs in the body and constant feedback loops about changing conditions and how to respond. Compared to the equilibrium's soothing hammock that swings back and forth, homeostasis is a bit like trying to pitch a tent in a hurricane.

Although the aim is something akin to equilibrium – a steady and regular set of conditions – the means of achieving it could not be more different.

Most of the time, we have to work as hard as our bodies to maintain at least some semblance of order in our lives. The seesaw that is our everyday life is constantly coming under different pressures on both sides – the things we do ourselves, and those that are done to us by others. Keeping at least a vague sense of balance is incredibly hard work and we have to constantly assess how one decision might have an equal and opposite reaction on our state of mind or well-being.

That means you have to make compromises with yourself and your sense of what the right decisions and life choices are. It's not possible to have it all. So while I want to go to the gym five times a week, sometimes a blocked nose and dry throat will be imploring me to take a rest. The body is saying one thing and the mind is demanding another. Even though I *always* want to do what the mind is telling me, I've had to learn that you need to listen to the body and let it guide how much exercise you can do on a given day. It's taken me until the age of twenty-six to fully appreciate this, after ten years of struggling.

Just as we can't defy the laws of thermodynamics, we can't stop the seesaw from moving. Disorder is in the system, as unavoidable as gravity. And if we're being honest, it's also something that many of us rely on, helping our lives to unfold organically. It's why we might be vague about when we're going to deliver a piece of work, or meet up with someone. It gives us the wriggle room we need, rather than establishing a firm commitment. By contrast, if I say I am going to meet someone 'midweek', then I definitely mean on Wednesday at midday. Why would I mean anything else but that?

It's important to recognize the reality that our lives are finely balanced, with inputs coming both from our own choices, and circumstances and decisions beyond our control. No decision we make is entirely isolated or costless in thermodynamic terms. Everything is a choice about how to expend energy, for what end and in whose favour. In turn, this is going to affect our ability to address all the other things on our seesaw.

There is never, or rarely, any such thing as total equilibrium for everything at once, because the factors are simply too many. Much like one of my favourite lines in Stephen Hawking's *A Brief History of Time*, which tells us that 'nothing is ever at rest'. The prospect of this caused many a sleepless night for me, as I waited up for the whole world to sleep along with me in phase.

But the more we realize that there is a seesaw, the more we can make conscious decisions that create a minimum level of balance and order. It's not perfect order, it's not total control, but it's as good as you're ever going to get.

There is something liberating about accepting the limits of how much order we can create in our lives. Once you've accepted that it's no more possible to live the perfectly planned life than it is for a sandcastle to resist the tide, it's easier to focus on the things you can control. There's enough on all our plates without having to deal with unrealistic expectations.

Having ironed those out, it's time to focus on what order you can create and how. The first step is to compromise with yourself: remember that the more precise your vision, the more energy it will take to achieve. So be sure, if you set yourself an exacting task, that it is going to be worth the effort.

There's no point tidying your room unless it is going to make either you or someone else feel better. And although it's human nature to want to do everything, it's better to prioritize the things that are likely to make the biggest difference, and to persuade yourself not to regret the ones you haven't got enough time or energy for.

After compromising with yourself, you have to do the same with other people. If you're sharing a house or an office with other people, then there are going to be competing visions of what the optimum temperature, layout and organization should be. Not everyone's can be achieved, but they can all be understood and taken into account. It might sound like a simple thing, but stepping back to appreciate how much energy it takes to do things – and understanding how that is rooted in the fundamentals of thermodynamics – can make a big difference. It frees us from the toxic assumption that we need to try to do everything at once, please everyone and meet all expectations. That isn't just unhelpful, it's also unachievable – and science has your back on this one. Compromise isn't giving up, it's adjusting to reality according to physics.

Living in a thermodynamically favourable way is all about making the right compromises. You need to understand your own sense of order and how you want things to be – and then be willing to break from it. You need to empathize with how other people see the world, and make accommodations without giving up on your own needs. And you have to embrace disorder, which isn't the same thing as surrendering to it.

Above all, you need to realize how unfavourable perfection is. Take it from me: being inflexible is one of the most exhausting things there is. By contrast, making conscious decisions about what you can and can't do in a given day or week, and feeling not a shred of guilt about it, is one of the most empowering. To embrace and toy with disorder defines what it is to be alive. Thank goodness we do, because life would be boring and stagnant otherwise — and energetically unfavourable for human evolution. Without disorder, you might as well be living as an inanimate object. A chair, perhaps (but not mine, since it is already taken).

4. How to feel the fear

Light, refraction and fear

It's 2.30 a.m., and my room is filled with thick darkness and ice-cold silence. There is no one here to realize how petrified I am. I want my mum, but she is at our family home, a forty-five-minute drive away. I can't stop feeling anxious about the colour orange and the texture of biscuits in my head, and the smell of new shampoo on my pillow. I can't sleep and I want to go home.

Night-time was always a moment of peak anxiety for me. My ADHD induced insomnia, while my ASD filled the waking hours with obsessive thoughts and fears. I would find myself caught in the middle: scared to sleep and afraid to be awake. I would often need my mum to move her pillow into my room and sleep on my floor, so I would feel safe enough to get through the night.

These night terrors were just one example of the fears that have followed me through life. There are the obvious anxiety triggers that still affect me today, like sudden loud noises, or large crowds of people. And then there are fears whose origin even now I struggle to understand. I can sip on a carrot-and-orange juice – my weekly treat – and wonder why this colour used to repulse me so much. Orange food, orange clothes, orange plastic seats, all once seemed like toxic or contagious substances to be avoided at all costs. This is part of how ASD works, creating instinctive and repulsive fears that can't be explained, but must be obeyed.

Fear is something we all have, and it's something we need, essential to our survival as a species. Without fear we have no scepticism, no caution, no check and balance on our impulses. But the opposite is also true. When all we can feel is fear, it becomes paralysing, leaving us unable to think clearly or make decisions at all. Your fears might be small ones, about a difficult meeting at work or admitting your feelings to someone. And they might be large: phobias you have always held, worries about major changes in your life, fears about ill health or financial issues. Whatever the case, fear accompanies us all whether we acknowledge it or not, and whatever size the dose. Unless we understand our fears, untangle their root causes and examine those issues rationally, we risk being controlled by the things that make us afraid, rather than taking control of them. Fear can be irrational but more often it's highly logical and reasonable; and our response to it must be the same.

With Asperger's, there are moments when all your thoughts and fears rush onto you like a beam of blinding light. You experience everything all at once and have no inherent ability to separate the different emotions, anxieties, impulses and stimuli. Another of my great fears was fire alarms, a terrifying noise that would send my senses running red hot as the noise reverberated through what felt like my entire body. Imagine a feeling of total physical dread. At school, while the other students would neatly form ranks like soldiers, I always had to run as far and fast away from the noise as possible.

At moments like these, I have to live in darkened rooms, with the blinds closed, noise-cancelling headphones on and quite possibly the safe canopy of my desk to sit under. This was, and is, my survival method. But it's not a way to live. I needed something that allowed me to get ahead of my fears, as well as to hide away from them. Because I have no in-built, unconscious filter, I knew I had to create my own: one that would allow me to cope with fear and function alongside it.

And, just as the feeling of fear felt like a blinding light, my study of photonics (photons being the quantum particles that make up light) helped me to realize that they could be broken apart in the same way a beam of light can be refracted, revealing its many different colours and frequencies. Our fears, which are never as singular or overwhelming as they sometimes feel, can be treated in exactly the same way. With the right filter, we can open up, understand and rationalize our fears – seeing them in a new light. So save your #nofilter for Instagram. In real life, we need all the filters we can get.

Why fear is like light

Shadow and light have always fascinated me. At home I had my favourite tree, in whose shade I would stand to feel safe. I have always needed areas of low light intensity as a protection against sensory overload.

But I also loved light and was enraptured by its properties. On the windowsill of my mum's bedroom she had placed a crystal oyster shell, which refracted the sunlight all around the room, revealing the natural treasures of the sun as it broke into its full spectrum of colours: piercing red at the top, and serene violet-blue at the bottom. Everything would come alive in that moment, at 7.30 every morning when I rushed to watch it, dreading the winter months when cloud would deprive me of the spectacle.

This was a moment of peace and wonder in a day that might be filled with all kinds of fear and anxiety. I instinctively knew that I needed a prism of my own for the thoughts and feelings that tangled in my head like spaghetti gone wrong. I had to separate my fears, understand everything they contained and unravel a sensation that was otherwise overwhelming.

I started off by reliving the most vivid parts of my day – sitting safely under my desk, naturally – and trying to associate each scenario with its most powerful emotion. What had I been feeling so strongly about and how had that contributed to the situation? As I plotted the map of my emotions, I kept coming back to those mornings watching the light refract through the oyster crystal. My anxiety attacks were like a beam of white sunlight – overpowering, impossible to look at directly, and something you can only turn (or run) away from. But within that sat a whole spectrum of emotions, some stronger and more immediate than others, all interplaying and tangling together to create fear.

Refraction was a perfect lens to help me understand and classify my fears because of my synaesthesia, a condition in which normally unconnected senses become linked to each other. For some people that means you can see sound or taste smells. For me, it has always meant that I feel colours as well as seeing them, and they all have a personality of their own. By seeing fears on the light spectrum, they became clearer to me, as well as more distinct.

You don't need to have synaesthesia to benefit from this perspective. Nor do you have to be someone who suffers from an abundance of crippling fears. We all have fears, at the front or back of our minds, and we all face moments when fear takes hold of us in ways we struggle to control. Has anyone ever told you in these moments to slow down, or pause for breath? Well, that is exactly what refraction achieves. When light passes from one substance to another, it will change speed. Because light travels less quickly through glass or water than air (both having a higher refractive index), the wave of light will slow down. In the classic prism example, like my mum's crystal, the light then disperses into its seven visible wavelengths: red, orange, yellow, green, blue, indigo, violet (plus the invisible: infrared and ultraviolet).

In other words, by slowing down the speed of the light wave, we are able to see it differently: in its full glory and

many colours. The prism effect gives us a new perspective, turning something that was singular and dazzling into a spectrum that is much clearer, and even more wondrous. If we want to understand our fears properly, we need to do exactly the same thing: look at them through a new lens, so we can see them differently, and change how we respond accordingly. In other words, we need to get on a wavelength with the things that make us afraid.

Getting on a wavelength

Refraction happens because light does not travel in straight lines, but in waves that oscillate and undulate dependent on their energetic differences at any one time. You have waves to transport light through space, and the same applies to sound waves, radio waves, X-rays and microwaves. They are all around us, but light waves are the only ones we can actually see.

Every wave, whether it's the one that allows a fishing boat to tune its long-wave radio (the only one that can reach it out at sea), or the one you use to cook a ready meal, has its own frequency. A high-frequency wave has tall peaks that occur close together, like a particularly spiky Toblerone bar. Its low-frequency cousin unwinds more gently, similar in form to a loosely coiled snake. The higher the frequency, the more energy is carried, but the less distance it can travel in a prism since it interacts with the atoms contained within, dissipating energy. The higher the frequency of light, the more it will bend upon contact with a medium of higher density than air (such as glass or water). The pace at which waves travel affects everything we see and hear around us: during a storm, you will see lightning flash before you hear thunder rumble, because light travels faster than sound (by about a million times through air, moving unencumbered, while sound interacts with the elements around it). In reality, both occurred at the same time.

When light refracts through a prism, we see the different colours because the higher refractive index of the glass has slowed down the waves to a speed that falls on the visible spectrum (waves that can be seen by the human eye). The refractive index simply quantifies the speed of light relative to the substance: a measure of optical density that tells us light travels more slowly through denser objects (so progressively slower through glass, which is denser than water, which is denser than air).

As the light makes this journey, we are able to see something that was previously invisible: that there are different colours within light, each with its own wavelength. Red is the longest, so travels the furthest, and is bent the least by the prism. Violet is the shortest, and refracts the most. These differing wavelengths explain why a rainbow will always have red at the top, bending around furthest, and violet at the bottom.

Wavelengths are important to our fear and light analogy for two reasons. Firstly, the initial sense of fear, the blinding white light, is not singular but actually contains many different emotions, triggers and root causes. And secondly, these are not all equal: like the different colours in light, our fears and anxieties have their own wavelengths, with varying intensity. Some will flare vividly over a short distance (which for me could be hearing a loud noise in the street), while others will maintain a less insistent, but more sustained, drumbeat in our heads (such as my fear of having to look people in the eye). The most powerful, insistent emotions are like high-frequency violets – intense and choppy – while the nagging feelings are more laid-back, low-frequency and long-lasting reds. And, just as happens at sea, sometimes different waves combine to create a tsunami of fear that you are powerless to prevent from crashing over you.

This was my biggest breakthrough in managing the fears that threatened to derail me. Anxiety is not a single, solid state resting in our heads, but a fluid entity that contains a multitude of different components. The concept of refraction can help us to separate these, untangling the different things that make us afraid, distinguishing between the high- and low-frequency triggers, and ultimately finding ways to manage them.

When I feel a panic coming on, I use the prism effect to diagnose the situation and try to avoid a full-on meltdown. Is it a high-frequency wave, a sensory trigger in my immediate surroundings, like someone accidentally brushing past me, shouting loudly or giggling at a high pitch? Or is it one of the low-frequency, constant thoughts that occupy me: fears about the future, getting ill, or whether my itchy jumper is going to give me psoriasis?

Am I experiencing an ADHD panic, where I feel nauseous because there isn't enough to stimulate me, or an ASD one, where there are too many options, my mind goes blank and I have to retreat into my cave? One feels like spinning outwards, a fairground ride that's going faster and faster; the other like spiralling inwards, detaching from the world and retreating into myself. Unless I understand which it is, and why, there is nothing I can do to prevent the gravity of a meltdown.

While there is no such thing as simply 'conquering' fears, it doesn't half help to understand what you're dealing with if you're going to manage them better. We need a prism. In fact, we need to be the prism.

Becoming a prism

With fear, our natural impulse is to try to make it smaller. We think that if we can compress our fears into the smallest possible box, locked away in the furthest recess of our minds, then we will be able to live free from its influence. But hoping that fears can be controlled in this way is akin to supposing the sun may stop rising one day. If something is giving us anxiety, it will continue to do so until we understand why that is, and what we can do about it. Denial is not an option, even if it is our instinctive first recourse.

I have tried this approach, giving up things I enjoyed: taking part in mud runs and extreme sports, buying posh jam at full price (which I do in defiance of a former boyfriend, who would only ever buy on offer), or even – the Holy Grail – falling in love. All are things I want but know will also make me afraid. But denial is worse than fear: a sort of mental constipation that traps you and eventually makes you hate yourself for being so safe. Being opaque in this way is no more sustainable than trying to hold your breath for ever: eventually your spirit will suffocate. It's better to risk feeling afraid than feel nothing at all – to make yourself transparent enough for the light to shine through.

Going cold turkey on my anxiety triggers didn't work, so I knew I needed a way to become transparent to them. It meant

I had to become the prism – not pushing fear down but opening it up, letting it shine through me and breaking it into its constituent parts so I could study it in detail, better understand its nature and ultimately cope with it.

Fear is an intangible, something that exists in our minds, and so our prism has to be a mental one too. It is about training the mind to filter fear, putting it through a virtual refractive prism, rather than letting the white light of anxiety cloud our ability to think rationally. This isn't something you can easily or quickly learn. A good way to start is by thinking about previous scenarios and trying to establish retrospectively what made you afraid. Usually there will be several contributing factors, so try to establish each in turn, and think about how they interacted with each other. Try to separate the things that caused the fear from those that simply compounded it. Recall which emotions were the most vivid and all-consuming. Tease apart the different strands until you have a full spectrum of emotions, spanning the high-frequency triggers and the low-frequency anxieties. By doing this, you map your fear, turning it from an intangible feeling of dread into something you can understand, explain and better navigate in future.

Over time, you get better at this until you are able to refract in real time, holding up your mental prism to fears as they happen, and hopefully finding a way through. It's not a foolproof method, but the more I do it, the better I get at coping with the fears and anxieties that crop up in my everyday life: from being willing to step out of the front door each morning, to navigating my morning commute and managing work and social situations. Instead of being a sponge for fear, soaking it up until I can absorb no more, I try to make myself a prism that can refract a high-intensity beam right through me.

I think about making myself into the densest possible prism, bringing together all my experiences of a particular event or fear together in one place. This gives me the processing power and mental density to reduce the speed the fear is travelling at – just like light through glass – minimizing the chances of being overwhelmed. I'm then back in control, able to study the new threads of colour and detail that have been opened up. This is hard when I get entangled in anxiety, and my head starts to spin like a disco ball in darkness, but it's only by becoming the denser prism, slowing the fear down, that I am able to cope with letting the light in.

For example, if someone asks me to look them in the eye, I will immediately feel a burst of short-wave, white-light fear. I need to work fast to avoid being overwhelmed by this instinctive, deep-seated fear, one that strikes at the core of my ASD self. Through the prism approach, I can separate out some of the waves within that white light: the longest, reddest one that represents a fundamental fear of human contact from the more immediate, violet wave that is my fear of someone's eyes burning through my public mask, seeing through the well-practised exterior and revealing my anxious core. Once these threads have been identified, I can start to rationalize with myself: yes, I don't relish this kind of human contact, but I know from experience that it won't actually hurt me. And no, this person probably isn't trying to look me in the eye to reveal anything: they're just trying to have a conversation. They're not going to find out simply by looking at me how much I have learned about them just through observation. Only once the different strands of the initial fear have been separated can I bring this kind of logic to bear. It's neither feasible nor sensible to try to rationalize the fear in its rawest, white-light state. First it has to go through the prism.

Building up this kind of density is also a good opportunity to bring all your thoughts together – allowing you to base decisions on patterns of accumulated data from past experience, not moments of panic or anxiety. By taking this approach we can improve not just our response to fear, but our overall decision-making process: a kind of high-intensity workout for the mind akin to the HIIT sessions that have become popular in the gym.

How, then, can we develop and hone our mental prisms to achieve this? It starts with behaving like one, and learning to be more transparent. Instead of being ashamed of the things that make us afraid, as if they represent weakness, we should be honest and open about them. We shouldn't be afraid of telling our friends and family about our deepest fears, and we certainly shouldn't feel embarrassed about sharing them – whether with a personal friend or a professional. Being transparent about the things that make us afraid is a necessary part of developing the prism mindset. It's what allows us to move on from the urge to compress our fears, and makes us ready to look at them through this new lens. Though it is important to note that this is a two-way process, since to open up you need to feel that it is safe to do so. This can be hard in a pressured environment where you have to perform, such as in professional life which drives us towards the robust, masculine facade of indifference – a very low refractive index if you ask me.

How to be transparent will depend on the individual. Just

as different materials all have their own refractive index, the speed at which light can travel through them, we must all develop our own comfort level. Some will find being transparent easier than others. I have always been something of an open book, showing and saying exactly how I feel. This transparency goes both ways. Lacking the realism filter, I might see a motivational poster on the London Underground that reads, 'Anything can happen to you', and think it means I am about to contract a deadly disease, or succumb to silent and immediate death. Have you ever thought you might be so open-minded that you overload yourself with anxiety every single day? Welcome, it is a hard life being so transparent that you quite literally scare yourself.

By contrast, you might be someone who is more inclined to hold on to their feelings, and less willing to share their fears, ironically from a fear of human judgement. But transparency, and honesty, are not avoidable if we want to get a handle on fear. Unless you learn to think and behave like a prism, you are going to struggle to mimic its wonderful ability to turn that anxiety beam into the beautiful, understandable and manageable wavelengths that created it. Being more open is the first step to managing our fears, and is the route to feeling alive again. And if that makes you afraid, then fantastic. You know exactly where to start.

Turning fear into inspiration

Having tussled with fear and anxiety throughout my life, I eventually came to an important realization. Instead of being one of my greatest liabilities, anxiety is actually one of my most important strengths. It allows me to accelerate possible outcomes in my head, reaching conclusions much more quickly (out of necessity, because there is so much data to process). The methods I've described in this chapter are an important part of how I turn the downward spiral of an anxiety attack into potential epiphanies – maximizing my processing power and capacity to bring together the different threads of my experiences and ideas.

These techniques allow me to function without being over-whelmed by fear, but that's not all. I also find something inspirational about looking into the white light of my fears. It's the same impulse that has always drawn humanity towards fire – a source of huge danger that also powered human evolution – and which explains why children will always try to look directly at the sun, even (perhaps especially) when they've been told it's dangerous.

Mental 'refraction' is a coping mechanism but also a cata-lyst. It disperses the blinding light of fear into something amazing: the colours of the rainbow. In the same way, the things that make us afraid also contain the ideas and the stimulus to inspire us. Our fears are full of rich thoughts that, when separated out in a way we can cope with, allow us to see ourselves and the world differently. To engage with the things that challenge us and make us afraid is also to come closer to the ones that make us feel alive – and which give us ideas about what to try next.

The denial approach to fear isn't just a bad way to feel less afraid. It also makes you miss out on things. If I had never faced up to my fear of looking people in the eye, I would have lost much of the human connection I value most of all, precisely because I find it hard to establish. I might not like the process of meeting someone's gaze, but I know that the end result will often be worth it.

By trying not to be afraid, you also limit your capacity to be creative, inspired and amazed by new or unexpected things. You stop learning, improving and evolving as a person. Fear is a part of us, and if we try to shut it down, we close off some part of ourselves as well. The better I have got at coping with my fears, the more I have realized how import-ant they are – and how much I would regret their absence.

Fear is a funny thing because, although I may have given the impression that it surrounds every aspect of my life (and that is to some extent true), in other ways I am comparatively fearless. For instance, telling people what I think has never been a problem for me, including those whom others may fear as authority figures. Human judgement just doesn't do it for me, since fearing your own kind doesn't make any sense.

Imagine the response when, at the age of ten, I told the head-master at my school to 'Mind your own business and stop reading my mail', after a letter I was writing to my parents in class had been confiscated. 'That letter wasn't meant for you, but for my parents. You shouldn't have opened it as it is none of your business.' That earned me a reprimand lasting several hours, but I wasn't afraid of the headmaster, his out-of-proportion ears or the callused, caveman-like finger that immediately directed me to his office. I believed that I was jus-tified, and I didn't fear him simply for being in a position of authority. The filter that makes many people wary of authority figures simply doesn't exist in my head – they have to earn it through actions and behaviour over time.

When it comes to fear, we all have our own particular

anxieties. I'm not afraid of the things that you probably are, but I can be terrified by things you wouldn't even notice, often deemed 'silly' or 'unnecessary' (which always makes it worse). Lacking the typical filters, I'm both overexposed to mundane things, and unaware of the many social conventions and norms that I haven't carefully taught myself through experience. I can be simultaneously overwhelmed thanks to my ASD, and then bored stiff courtesy of ADHD. I might get completely thrown off by a change of routine in a gym class, but very calm if I learn a family member or friend has cancer (which makes me a bad exercise partner, but an excellent listener and therapist). When you actually have to live with #nofilter, it can be a disorientating experience, but it's also a true representation of the strengths of neurodiversity, and the contrasting skills it allows us to bring to the table.

Whether you are someone with few filters or many, there is one I believe we all need regardless, and that is the prism perspective on fear. We need the dispersal effect of the prism to turn fear from something overwhelming into a force we can control, and ultimately embrace. It's important to think about being in control of fear, rather than simply banishing it from our lives. Fear is something we need, and which can be part of how we inspire and motivate ourselves. When we are afraid, we are also reminded of what is valuable in our lives, and of our human instinct to protect the people and things we love.

If we try to lock fear away in mental boxes, then we lose all the advantages and keep all of the costs. By contrast, embracing our fears and putting them through the mental prism helps us to turn fear into an asset we can manage, like harnessing electricity from tidal waves. I know there will never be a day of my life when I don't feel afraid. But I also know that it's thanks to fear that I really feel alive. Fear isn't something we need to 'shine a light on'. It is the light, one we can all learn better how to live with, and even benefit from. It's why I see the fear that my ASD instils in me not as a problem to solve, but as a blinding privilege to take advantage of.

5. How to find harmony

Wave theory, harmonic motion and finding
your resonant frequency

For any parent, the combination of long, grey afternoons and bored children must be one of the toughest tests. And when you make it a bored, ADHD child, the task becomes twice as hard. My dad was always a brilliant parent, and never more so than in his endless resourcefulness at finding things for me to do.

He knew that giving me something to experiment with was the best way of keeping boredom at bay. One of these experiments would be regularly conducted at the river near our home, where we were saved from fractious weekend and summer holiday afternoons by the simple, timeless fascination of pebble skimming.

I'm willing to bet that everyone reading this has tried their hand at it, quite literally. Like me, you've probably seen endless attempts sink on impact; failing to achieve the ultimate satisfaction of a stone that bounces joyfully across the water, dispersing a trail of ripples in its wake. As my dad would say to us, it's the movement of the stone that brings the river to life – nothing happens if nothing happens.

My obsessive, scientific tendencies led me to spend hours searching for exactly the right kind of stone to skim: one with the flat surface that could maximize my ability to create magic ripples across the water. Yet on many attempts I caused nothing other than a dismal splash. It was only on rare occasions that my throw achieved perfection, the pebble cresting the water as if it had been born to, bringing excitement to the otherwise meandering river. When this happens there is a perfect harmony between the surface and the object disturbing it, exerting competing forces that allow the stone to continue leaping forwards on its journey. It was an early lesson in the beauty and significance of our actions and the often marginal difference between creating a positive or negative impact.

Across the still surfaces of our lives, new people and circumstances are constantly casting their pebbles. Some will sink painfully, others without us even noticing, but, as even the amateur stone skipper can occasionally get their pebble to bounce, every once in a while something hits us at just the right angle. We meet a person who will change our lives for the better, have an idea that allows us to see the world in a different way, or read something that alters our perspective. The stone of a new encounter skips, spreading its ripples throughout our consciousness.

I believe such moments are not accidents, or at least they don't have to be. There is a science to these synchronicities just as there is a technique for getting the pebble to bounce across the river. Waves, such as those that the stone creates as it passes over the water, are fundamental to our understanding of motion, dynamics and mechanics. Some of the most important ideas in physics have been derived from the study of how waves move, oscillate and interact with each other.

At this level, humans are not so different from light, sounds and tides. Our personalities, our relationships and our moods oscillate as waves – undulating upwards and downwards, altered by the parallel and opposing waves they will encounter. The mechanics behind oscillation, harmonic motion and wave theory help us to understand the vicissitudes of our own temperaments – and how to match that inner tune to those of others, avoiding disharmony and getting in tune with the people, places and situations that define our lives.

Harmonic motion and amplitude

It might have been a long time ago, but I'm willing to bet you can remember going to the park as a child and racing to the playground swings; or perhaps you have a young child now who is always pleading to be pushed higher. There is something unforgettable about the rhythmical movement of the swing, soaring to the top of its arc and back down again, again and again. It's joyous, liberating and above all reliable. You are protected on the one end by the centripetal forces (ones that make objects follow a curved path, back to the central point) that stop you from flying off into the distance, and at the other by the reassuring presence of a parent or sibling to 'catch' you at the bottom.

The humble playground swing isn't just a childhood favourite, or a plastic seat and some metal chains attached to a frame. It's also an oscillator, a system which experiences a repeated pattern of motion between two fixed positions; and an example of simple harmonic motion, where the force bringing you back down is equivalent to the one that 'displaced' (pushed) you from equilibrium in the first place. Oscillators – other examples of which are springs and the pendulums in grandfather clocks – are an important experimental tool in physics, used to model systems far more complex than you will find in the playground. They can also help us to understand the principles of wave theory and mechanical physics that put our personalities and relationships in a new light.

Just like people, oscillators can be predictable and they can be unpredictable. They have an expected path to follow, but one that is subject to change depending on the external forces being exerted on them: in the swing example, you might drag your feet on the ground, creating friction, or propel yourself outwards, revving into a salient rhythm. And, just as our personalities can be subdued or extreme, oscillators can have amplitudes (peaks and troughs of their arc) that are spiky or shallow.

To understand what the playground swing has to teach us, we need to represent the oscillation as a wave pattern.

In this diagram, the horizontal line is the equilibrium

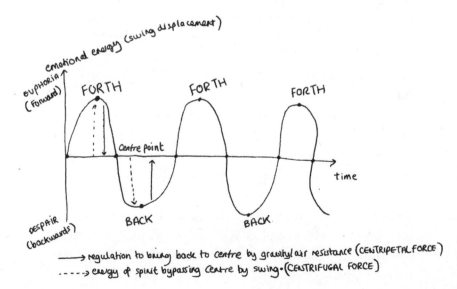

point where we start, sitting still on the seat, waiting to be pushed. The highest and lowest points of the waves are the amplitude – essentially the 'size' of our swinging arc. Every wave, whether it's one you see at the beach or the music play-ing in your headphones, has an amplitude that is reflected in the size of what we see or the noise of what we hear. The size of the amplitude is proportional to the amount of energy it is carrying: the force with which we were propelled on our swing. Our wave also has a frequency, the rate at which we are oscillating, which essentially measures speed (when the wave peaks are coming closer together, you are moving faster).

In the last chapter, we looked at waves in terms of light and refraction. Here I want us to start thinking about ourselves through the same prism. Because just as, on the swing, we are quite literally riding a wave of simple harmonic motion, our lives and personalities have their own inherent wave patterns.

Think of the people you know who always seem to be in control of their emotions, never get overtly bothered by problems and are essentially 'steady'. That is a low-amplitude personality, one which never departs too far from its resting equilibrium. Neither the emotional forces propelling this person nor those dragging them back are ever too great. This is a slow, steady, smooth ride on the swing: no sudden jerks or motion sickness.

By contrast, a high-amplitude person is someone with more energy to burn, whose emotional peaks and troughs are more extreme, and who is also probably travelling faster – at a higher frequency. This is the kind of ride on the swing that can make you feel sick, as the soaring highs are tempered by unsteady downswings and sudden jolts of force when you're not expecting it. I am, of course, describing myself here and especially my ADHD.

ADHDs, to use another classic example of simple harmonic motion, are the most coiled of coiled springs. We oscillate vigorously, impatiently and in a way that can often intimidate or confuse others. Our energy is greater and that results in a larger, less stable and more dramatic amplitude. Because when the force propelling you is much greater, the laws of motion tell us that its opposite – the one bringing you back down – is going to be bigger too. It can be exhausting to live with these amplified 'up' and 'down' forces constantly ramping your oscillator from one extreme to the next.

The frenzied ADHD wave pattern reflects that we have more energy to squeeze into the parameters of our day, requiring us to operate at a very high frequency, one that leads to impulsive behaviour and short attention spans. The energy has to go somewhere, so our oscillator cranks up to a higher frequency and amplitude to create outlets. These impulses respect few conventions and explain why I have sometimes gone outside at twilight, in my PJs, for no other reason than to jump on the trampoline: the coil of my own spring as tight as those propelling me up towards the setting sun.

Whether you are a high- or low-amplitude personality – and both have their strengths and weaknesses – it's important to understand the rate at which your temperament swings. It's only my relatively recent diagnosis of ADHD that has allowed me to properly appreciate the way I function, and make helpful adjustments: from where I live to who I hang out with. To make our progress through life as natural as the 'good' ride on the swing, we need an understanding both of our own amplitude and that of the people around us.

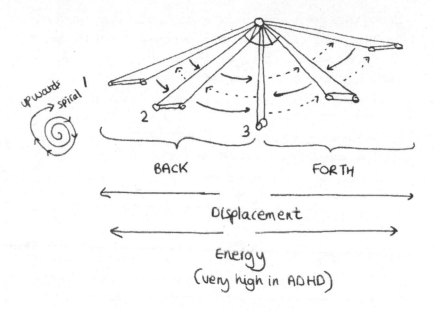

Only then can we hope to find an energetic harmony within ourselves – the basis of being able to find the same with other people.

Constructive interference and resonance

This appreciation of amplitude is important because, when it comes to oscillators and waves, there is no such thing as an isolated existence. A harmonic oscillator doesn't live in the perfect, frictionless world that might allow a ball on a spring to go back and forth to eternity. So too is the child on the swing affected by everything from wind resistance to the timing of the person waiting for them at the bottom.

It's the same across the rest of our lives. The wave of our

personality doesn't simply unwind happily through space and time, merrily following an uninterrupted path. Instead it encounters and interacts with other waves – ones which can materially change its form, pace and direction.

There are two concepts that help to illustrate this process, and demonstrate the kind of external circumstances we must prepare and adjust for. They are interference – what happens when two waves come together – and resonance, the impact of external force on a wave pattern.

Interference

Interference is about the combination of waves and the effects they can have on each other. When two waves interact, the one that is created is said to be the result of the 'superposition': the sum of their two amplitudes at the moment they overlap.

This can produce one of two results. In *constructive* interference, two waves whose amplitudes are in sync combine to form an even bigger wave. This is much like when you sit at the beach and see waves rolling over each other towards you, the combined peak getting bigger and bigger. It's when you meet someone who allows you to be an even better version of yourself.

But waves – whether light, sound or tidal – are not always in sync. And when two waves intersect at the moment when one is at its peak and the other near its trough, the opposite happens. This is *destructive* interference, where two competing waves effectively cancel each other out, returning a situation to equilibrium. The positive amplitude of the peak and the negative of the trough add together to form a superposition of zero: one where nothing is seen or heard. People who destructively interfere are those who sap your energy and joy, neutralizing it on impact through their negativity.

An example is one of my favourite and most important possessions: noise-cancelling headphones. I never leave the house without my pair, and the shield they provide against loud conversations in cafés, ambulance sirens in the street and people shouting from car windows. It's thanks to destructive interference that I can safely navigate places that would otherwise be impossible: the headphones create a sound wave designed to be sufficiently out of phase with surrounding waves that my ears hear nothing: the equilibrium of two waves whose amplitudes have added up to zero.

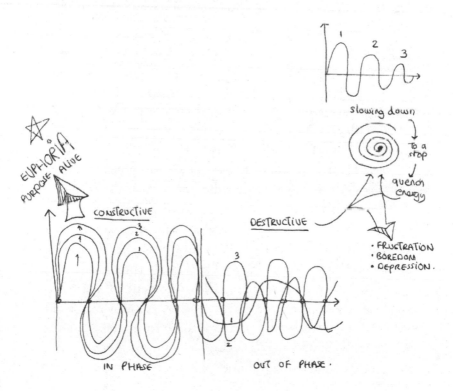

Waves, then, can either amplify or nullify each other. They can combine to create something greater than they could be alone. Or they can clash to the point of equilibrium: where the noise falls quiet, the light goes out and the energy dwindles away. And they can do any kind of variation on the spectrum in between. It's not the inherent properties of the waves that make them disharmonious, but the point at which they meet that matters most. Two waves that could be constructive when in sync become destructive when out of phase. As with anything in life, timing is key.

Resonance

Timing is also essential when it comes to resonance, another essential concept in understanding the life lessons of wave theory and harmonic motion. This refers to the 'resonant' frequency at which a wave system most naturally operates – the noise made by a wine glass when you tap it with your fingernail or the natural arc of our playground swing in motion. The power of resonance is what happens when another force is applied: its impact on the object in question depends much more on whether it shares a frequency than how powerful it is. In other words, a relatively small push on the swing at just the right moment in its arc – one that aligns with its resonant frequency – will be much more effective than a massive shove that is badly timed, before or after you have reached the critical moment of the backswing. Or with the wine glass, another sound wave on the same frequency can exert enough force to shatter it, in the way a really loud noise (and higher amplitude wave) on a different frequency would not.

What resonance and interference show us is that synchronicity is what really makes the difference in nature. Being 'in phase' is often more impactful than being incredibly powerful. The same is true of human personalities and how they work together — or don't. If you've ever said about a new friend or love interest that 'We just clicked', then you know what constructive interference feels like on a human level: a relationship that is so much more fun, energizing and life-giving because two personalities lift each other to a level that neither could reach unassisted. On the other hand, you have probably also experienced the opposite — when someone who is *not* in sync with you actively seems to drain your energy and enjoyment. Just as an annoying noise can be 'cancelled' to equilibrium by another, out-of-phase wave, our spirit and personality can be neutralized by the wrong kind of people in our lives. These destructive interferers tire us out, make us feel bad about ourselves and make it impossible to enjoy anything.

As someone who has often felt on a different wavelength from the rest of the world, understanding frequency, interference and resonance has helped me to better manage the ups and downs of social interaction. They help me work out who I should be surrounding myself with, the people and situations that lift me up or dampen me down dangerously, and how to make the most of my somewhat extreme personality: as someone who I know can be 'difficult' for other people, but who also has a lot of love to give. Next I want to look at how we can all use these concepts to get on a wavelength with the people in our lives: making the most of the relationships that enrich our lives, and steering away from those which detract from them.

Getting on a wavelength

Ever since early childhood, I have always felt a sense of alien-
ation from other people, that there are things they understand
which I don't, and feelings they feel which I can't. It has taken
most of my life to realize that I am not just on a different
metaphorical wavelength from most of my fellow humans,
but in a literal neurological sense.

Just as sound and light travel through the air at different
frequencies, so too do the motions of waves in our heads:
from gentle, low-frequency waves when we are sleeping –
referred to as delta and theta waves – to slightly higher-pitched
ones – alphas – when we are relaxing and doing nothing in
particular, to the high-frequency ones that pulse when we are
actively engaged in a task that requires us to be alert and
focused – beta and gamma waves. Depending on the situation
we find ourselves in, our brainwaves can create a soothing
neurological harmony, or a percussive drumbeat.

With ADHD, your brain is often on the wrong wavelength for a particular situation: out of rhythm with your
peers, in a neurochemical silent disco all of its own. Studies
suggest that an ADHD brain is likely to be stuck in theta
mode when the task at hand requires the more active beta. The
net result is that your sense of time and space collapses into an
undefined mess, like living underwater. The world is moving at
one pace and your brain is going at another. This can mean a
combination of anxiety, and days in which all your energy
is consumed by trying to find your bearings. Managing my mind
is like trying to look after a roomful of toddlers: fitfully restful
but all too often crying, screaming or laughing uncontrollably.

To understand what this feels like, imagine you are trying to drive a Ferrari down a busy high street. The pace your brain wants to go at simply isn't compatible with the environment it finds itself in. You are jumping around from one thing to the next, constantly pumping the mental accelerator, while all around you there are pedestrians, other vehicles and traffic lights. Your brain wants to go faster, faster, faster, but you keep hitting the traffic-calming measures of everyday life: remembering your keys, getting to work on time, eating lunch, being nice to people. It's hard work.

Having ADHD doesn't just make it hard to concentrate for long periods of time. It also makes you highly impulsive and moody, and prone to swing from euphoria one moment to energetic despair the next. Your attention sways wildly too, a weathervane in a perpetual gust of distractions.

I might go into the kitchen with the innocent intention of making a cup of tea, only to pick up an interesting book while it brews, forget all about the tea while I find a notepad to jot down notes, suddenly decide to go for a walk to buy groceries, return home with a pack of chewing gum to calm my anxiety, realize the abandoned tea has stained my mug, put on a pair of Marigolds to wash it, and forget about it again because I want to post a picture of myself wearing the gloves on Instagram. It's a lot of effort for a cup of tea you're never going to drink.

My ADHD brain is a high-frequency, high-amplitude beast: one in which both the peaks and the troughs can be quite extreme.

That means I have to be careful about how I interact with other wavelengths: both for my own good and for everyone else's. I know I can be a *tad* dramatic at times, because sometimes my inherent energy and enthusiasm spill over into what feels to others like terseness, excessive directness or an overly emotional response.

'You're too much, Millie', was something I used to hear a lot. Today, at work, I have to make sure that in my eagerness to make a point or present an idea I am not accidentally about to offend someone by undercutting or speaking over them. I also have to remind myself that enthusiasm isn't the cardinal sin and I must keep expressing my spirit. After all, everyone loves it when a burst of energy (or a tray of baked goods) lifts the morale of the dry office environment. Managing my relationship with the outside world is an uncertain business, like trying to tune into a malfunctioning radio: sometimes you hit the perfect frequency, but often you are treated instead to a burst of painful static.

At the same time, I have to be careful to look after myself and make sure the other wavelengths in my life (which can be people, places and situations) are more constructive than destructive. I have experienced depression several times in my life, and I believe the root cause has been finding myself in circumstances where my natural amplitude was so out of sync with the context that it became impossible to bridge the gap. I was expending more and more energy to try to keep going, most of it futile. My high-amplitude spirit often fell into a painful silence. And the more this differential between me and my environment became clear, the lonelier and more isolated I became – frustrated at my own futility and suspicious of my natural personality and needs. This silent logic was the root of my depression, making me question myself as much as my environment: fearing that every action I undertook would be a wrong turn. When I could function at all, it was in a dampened and depressive state, burying my real self and putting on the mask of normality to interact with people, something that took up even more energy and, over time, gradually made things even worse.

The first time I experienced this was during my PhD, when I had reached the stage where you have to check every comma of something you have already rewritten four times. My mind was buzzing with so many ideas for things I wanted to do, but I was being forced to push myself to the level of lowest amplitude. As time went on, and I struggled more and more with having to limit my imagination and focus, I began to lose all my energy and enthusiasm. I would get out of bed an hour later each day. I struggled to function at all, let alone focus on the demands on my work.

Something similar happened more recently, when I moved to Slough for work. I took the job thinking that a 9 to 5 on an industrial estate might prove to be something of a restful 'gap year'. Instead, the steady accumulation of coffee-shop loyalty cards and the unnervingly slow automatic doors were anything but. I was getting none of the stimulation I needed, and again felt badly out of sync with my environment. I could almost feel my energy sinking into the beige carpets and leaching into the fluorescent lights. In the end, some good came of this, because at the age of twenty-six, as well as making some great friends, I got a formal diagnosis of ADHD, something that helped me to understand better the sense of asymmetry I had often felt with my planet and species.

What those two episodes taught me, reinforced by friendships and relationships both good and bad, is that we must be sensitive about the gap between amplitudes – whether those are the wavelengths of two people, or your own and the place you are living or working in. Because I have such a naturally high amplitude and energy, oscillating from getting deliriously excited about a new idea to becoming overwhelmingly anxious about an odd smell or a colour I dislike, I struggle to deal with people and situations at a completely different amplitude – where I expect others to be indifferent or even hostile, and am constantly overthinking how I can (or don't) fit in.

If someone else's natural level is so much more relaxed and easygoing than mine, then they are going to frustrate me,

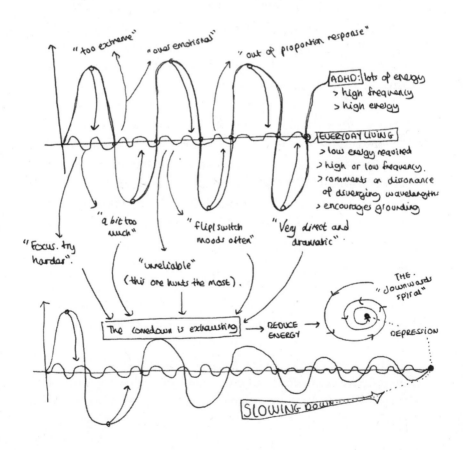

while I almost always end up feeling a thread of guilt, since I can often freak them out by being too intense. This is why, apart from people who I know 'get' me, I will often choose to spend time on my own rather than with others where I know we will end up in destructive interference. I know from experience how tiring it is trying to conform to the environmental and social norm. For me, that effort just isn't worthwhile.

By contrast, just as being out of sync with our environment is exhausting and dispiriting, finding others on the same resonant frequency is one of the most wonderful things in life. Like the perfectly timed push on the swing, a friend, partner or work colleague we are in sync with can have a massively positive effect with minimum effort: a comment, joke, gesture or message of only a few words. Just as the wrong people can sink our spirit and mood, the right ones can send it soaring. The magic of constructive interference helps you and your partners in crime – socially, romantically, professionally – to live your best lives together, more than you ever could apart. Because I am aware of my difference from most people, my awareness of the relatively small number I do feel in tune with is heightened. I rush to these beacons of constructive interference: those who can match, complement and sometimes gently temper the bursts of my personality.

You know these people when you find them – there is the intangible sense of spark, connection and togetherness that's often talked about to explain why people become friends or lovers. When someone says about a new acquaintance, 'I feel like I've known them my whole life', it's not an entirely irrational statement. Although you might not actually know that person, your wavelengths of personality and temperament have enough in common that your behaviour, assumptions and preferences are unconsciously aligned, even before you so much as shake hands. You have been walking large parts of the same path for years before your lives happen to converge.

When I talk about being in phase with people, that doesn't mean you have to completely share an amplitude. Being 'on a wavelength' with someone doesn't require that your waves exactly mirror each other. In fact, it's probably not a good thing if they do. Just as musical harmony requires different notes to come together, human harmony is about personalities that are in sync: not so different that the gap becomes too difficult to bridge, but not so similar either that you don't provide an effective check and balance on each other. Like the stone dancing across the water, two objects – or people – don't have to be especially similar to create something beautiful together. It's all about the angle – and the timing – of the interaction.

My best friends are those who I know will help me to moderate my most excitable moods and to lift me from my deepest troughs; and for whom I can do the same in return. As the diagram on the opposite page suggests, it's about your wave patterns sharing enough of the journey, while retaining your individuality and ability to complement each other. We need the challenge and potential for change that contrasting waves (personalities) offer us – this creates the potential for exploration, the key ingredient to any scientific experiment and every life well lived. But this kind of diversity only works within certain boundaries – enough that two people can adapt to their contrasting frequencies, ride out the natural disturbances, and benefit

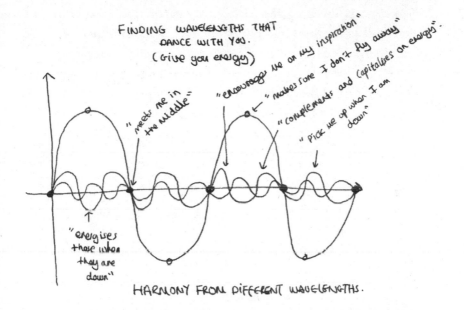

FINDING WAVELENGTHS THAT DANCE WITH YOU. (Give you energy)

"meets me in the middle"

"encourages me on my inspiration"

"makes sure I don't fly away"

"complements and capitalises on energy"

"Pick me up when I am down"

"energises those when they are down"

HARMONY FROM DIFFERENT WAVELENGTHS.

from their differences rather than being overwhelmed by them.

To extend the musical analogy, our lives are a little like playing in an orchestra without a conductor. We are all sawing away on our own instruments, hoping to find harmony with those around us, while others play their own tunes, often discordant to our own. Because there is no conductor to get everyone playing in sync, we have to listen out ourselves for those who are likely to be harmonious counterparts, and the people who we are always bound to clash with, however hard we try. What we should be listening out for is the resonance: the people, working environments and places to live that will uplift us almost by definition — just by being themselves, because they are aligned with our resonant frequency. That resonance is something most of us spend our whole lives looking for – seeking the friends, life partner, job and home that will give us an intrinsic sense of peace, fulfilment and happiness. That search must start with an understanding of your own wavelength, and developing empathy for those of others. On life's pendulum, we all have to find our own rhythm, and the people who can help us dance to it.

6.   How not to follow the crowd

Molecular dynamics, conformity and individuality

I have always been fascinated by how things and people move. Aged five, I would sit watching the dust particles float across the sunlight that beamed through my bedroom window. I was mesmerized by their abundance and how the particles moved in phases, most together while a few always seemed to go astray. I would sit in the morning sunlight, eyes closed, feeling the warmth on my face and counting how many particles I could feel landing on my cheeks. In fact, I was only allowed to do this for fifteen minutes a day, because I enjoyed it so much that I would otherwise have happily sat there all day, basking in the dusty sunlight.

The movement enraptured me, and so did the sense of magnitude: that we as humans could be so ultimately insignificant in number to other things we could hardly see or understand. At this point in my life, before I had learned any biochemistry, the smallest thing I understood was one I had just been taught in school: the full stop. This was my proxy for what I would later come to know as the atom. Understanding these would surely hold the secret to the clouds of dust that I bathed myself in every morning.

As I sat daydreaming, my mum's voice floated up the stairs. 'Millie! I'm not going to ask you again, what do you want on your toast?' Having swaggered downstairs in my fake glasses

(I still held a candle for Elton John at this point), I blurted out the much more important question on my mind. 'Mum, how many full stops are there in the world?' Her eyebrows furrowed in laughter. 'That's a yes to Vegemite then, I take it?'

I never got a satisfactory answer about the full stops, but ever since then I have been someone who observes and analyses how the world around me moves. I would sit in a café, pretending to read a book, but actually watching how people move around and behave: on their own paths and relative to each other. What was predictable and what was random? How much could I rely on the dynamic behaviour of others when it came to finding my own anxiety-ridden path through the crowd?

I watched and I read: Thomas Hobbes on the nature of man, Adolphe Quetelet on *l'homme moyen* (an average person whose behaviour would represent the mean of the population as a whole). I played *Civilization V*, as a way of simulating how different human decisions mapped out on a grand scale. And, every time I got on the train or sat in the playground at school, I observed and learned more about how people behave relative to each other, and the patterns of human movement.

The question I was trying to answer was a fundamental one. Is our behaviour essentially individual or conformist? Do we move according to our own rhythm, or to the drumbeat of the crowd? Are we one of the particles of dust that make up the cloud, or one of the outliers? Unlike Hobbes – although I think I could definitely work the ruff collar – my motivation wasn't philosophical. For me, this was a deeply practical question. Unless I could project with some degree of certainty how the people around me were likely to behave, I would never feel safe among them (or indeed anywhere near them). Before I could summon the courage to attempt a journey through the scary, smelly crowds of people who pack every shop, pavement and train platform, I had to understand their norms. I needed to study them so I could look after and reassure myself. Otherwise it would be back to where so many childhood outings ended up, hiding in our car being comforted by my sister, a coat over my head to block out the noise and light.

It was Lydia who first made me think about the crowds that scared me in terms of a game. Turning a busy street into a kind of human *Tetris* would allow me to make light of the situation, and put my scientist's hat (and coat) on. I could turn something that made me afraid into one I actually enjoyed: a theoretical problem to be studied.

All of which has meant that, although crowds still rank among my greatest fears, watching people is one of the great pleasures of my life. I can get more entertainment tracking the vagaries of pedestrians crossing the road than I can from a whole series on Netflix. It's my equivalent of cavemen sitting around the fire, watching it burn. Call me boring, but there is nothing tedious about human behaviour, even in the most mundane situations. In fact, like the classical elements of earth, wind, water and fire that so fascinated ancient scientists, there is unpredictability and intrigue at every turn. These everyday narratives might seem slow, but when you start to look at all the branching stories happening right around you, there is enough to sate the appetite of even the most impatient person. Much more dramatic than a scripted TV show or film – which strings its narrative across a predictable arrow of time – can ever hope to be.

I believe that everyone can benefit from what I've learned

through this process, even if walking down the street is something you can do without having to think twice (in my case twenty times). The conflict of individual and collective applies to us all on some level. When it comes to setting the course of our lives, we all face decisions about the things we want versus what society expects of us, or compels us to do. Almost every major decision we make has both personal and communal motivations, and sometimes they pull us in opposing directions. Balancing individual needs and collective demands can be one of our greatest challenges.

We need to understand the context of our lives, and the behaviour of the people and environment around us, if we are to plot an individual course with confidence. Are our behaviours normal, and do they need to be? Can you be an outlier without becoming ostracized from the common pool? Does it matter if we want and need different things from the people around us? To learn about ourselves, we have to look outwards and study the movement of the crowd through both space and time.

Crowds and consensus

Is a crowd defined by the behaviour of the collective, or by the many individuals who are a part of it? Or for my purposes, as I sought to plot a path that would avoid unnecessary interaction, should I be looking at the individual people or the pattern of collective movement as my guide?

My inclination was to start from the bottom, with movement at the molecular level that I had read about in my chemistry books. Perhaps this could be scaled to a human level, modelling a predicted trajectory for each individual person, just as I would track a molecule moving through a force field. This led me to observe how people moved differently, some stepping aside out of politeness or kindness, others being more assertive and adamant about their right of way, rushing because they were busy or wanted to look as though they were. There were fast people and slow ones; bulky ones and smaller, more agile bodies. A diverse mix: like the atoms that have created them in all their forms.

What I quickly discovered is that trying to account for the movement of every single individual is simply not possible. This is what I instinctively want to do on every commute, but it leaves me exhausted and in dire need of a cat nap. Like trying to count dust particles, you will soon find that you run out of time, patience or energy.

Trying to measure things at the individual level isn't just impractical. It's also unhelpful scientifically because people, like particles, don't act entirely independently. We are part of a system, a wider environment of tangible and intangible components – from other people to inanimate objects, the climate and social conventions. We participate in the system, and we are also in many ways shaped by it. Consciously or otherwise, we observe and absorb the behaviour of the people around us. It conditions our assumptions and indirectly helps determine our actions. A flock of birds can change direction in a matter of seconds, because of how thousands respond to and anticipate the movements of just a handful. At different speeds, the same happens to us as we assess which way the person walking towards us on a pavement is going to go, or how people are likely to react to a major life decision.

The existence of the system gives us something else to

measure: something more feasible. And while you might assume that analysing a system can't tell us much about the behaviour of its components, kinetic and particle theory would disagree with you. Because while an individual might display apparently random and unpredictable behaviour, systems as a whole are more reliable actors and more valuable witnesses. They have been my starting point to understanding how to manage my own motion relative to everyone else's.

The key concept here is Brownian motion, the theory that explains how particles move around. It shows that particles suspended in a fluid (which can be liquid or gas) move around randomly as they collide with the other molecules within it. This is the molecules we can't see (without a microscope) pushing around the ones we can, through sheer force of numbers – the pace and direction of the movement determined by the unique factors of the local environment. Brownian motion shows us that, while it's important to focus on the big picture, we have to look at smaller-scale events to understand both how and why change happens. This is true whether you are looking at a decision in your own life, the movement of a crowd or the evolution of an economy. What is happening at the smallest conceivable level, when aggregated together, makes a big difference to the overall landscape.

The theory – which played an important part in establishing the existence of what we now know as atoms and molecules – was inspired by the Scottish scientist Robert Brown, who wanted to explain how pollen could move across the surface of an apparently still lake. Its chronology dates back to the Roman philosopher Lucretius, who wrote about how dust particles move through light, two millennia before the same sight captivated five-year-old me.

But although the essence of Brownian motion is about unpredictable movement – the progress of each particle is even known as a random walk – that's not the whole story. Microscopically, every particle is doing its own sweet thing, buffeted this way and that by the liquid or gas molecules that surround it. But change your perspective to the macroscopic – the big picture – and you see something quite different. Through this zoomed-out lens, randomness starts to give way to a pattern. The collisions between molecules are unpredictable, but their overall effect is the opposite. Via Brownian motion, the particles in question will disperse about evenly across the surrounding fluid. This can be seen through diffusion, by which particles move from high-concentration to low-concentration areas, until they are evenly distributed (the reason you can smell baking right through your house, even though it is only actually happening in the oven).

Like the pollen or the dust, as individuals we follow an unpredictable path – one conditioned by how we interact

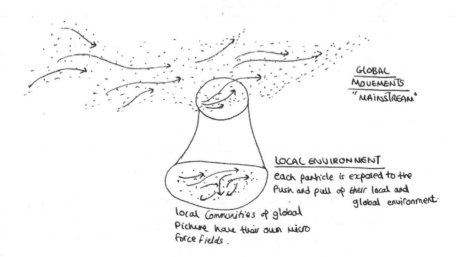

GLOBAL
MOVEMENTS
"MAINSTREAM"

LOCAL ENVIRONMENT
each particle is exposed to the
push and pull of their local and
global environment.

local communities of global
picture have their own micro
force fields.

with our surrounding environment. But when all those paths are modelled and viewed together (thanks to a handy technique known as multidimensional scaling), the direction of travel becomes clear, allowing us to see what is going on overall.

This realization allowed me to take a formulaic approach to navigating myself around busy city centres and streets. Using Newton's second law – force = mass × acceleration – I could predict the likely path of traffic as long as I knew the relative proportions of the different elements – people – and perhaps some context about the time of day and where most of them were heading. So the town centre on a Saturday, with lots of heavier atoms headed towards a rugby match, was very different from how it was during the school run, which had its own molecular peculiarities. Each distinct environment was created by the different molecules involved, their movement and interactions: something that can be studied in the same way as molecular dynamics, the science of how molecules move through a force field over time.

For every place I would regularly visit, and at the times I was likely to be there, I used Newton's law to help me determine a formula for how people would move. In fact, this was one of the reasons I have always wanted to make myself physically small, so my own mass has as little effect as possible on the overall experiment (the observer effect, in which you seek to minimize the human error or influence of observation on the natural behaviour of your sample).

By understanding and modelling consensus behaviours I gradually managed to counteract some of the innate fear I felt in the presence of crowds, with a sense of certainty about how they would behave. My anxiety started to give way to waves of euphoria, releasing me from the chains of the attacks that would previously hit every time I stepped outside. Now I had a compass and a map to navigate a situation that once routinely sent me into meltdown. Brownian motion had convinced me that there was enough certainty to make it safe. I could plot my own path.

Crowds and individuality

If studying crowds taught me something about conformity, even more important was what I learned about individuality. Although modelling systems can demonstrate the existence of consensus behaviours, it by no means follows that we as humans are homogeneous. In fact, one of the most irrational human beliefs is that there is such a thing as the rational or normal way to do anything. When you have ASD, you quickly realize that people invoking the 'normal' is usually a thin veil for fear or prejudice.

Looking at the crowd through another lens shows that, just as there are patterns to be identified across individual behaviour, there is a significant level of variance within the consensus.

This is where ergodic theory, a mathematical idea used to study dynamic systems over a period of time, can help us out. It holds that any statistically significant sample of a given system will display average properties of the whole, since any of these microstates is theoretically as probable as any other to be occurring elsewhere. A different state, somewhere else in the system, is no more or less likely than the one you are currently observing. In other words, my 'normal' is just as likely to be indicative as yours within a stochastic (randomly occurring) process of a suitable scale, observed for a sufficient amount of time. Take the dust clouds that once used to fascinate me. Any individual particle is actually an indicative microcosm of the whole system, displaying average behaviour in both its conformity and randomness. It's no more or less normal for a particle to be an outlier than part of the main grouping over the entire lifetime of the system. Its range of movements will always be representative of the whole, as long as you track its behaviour fully across space and time. In the same way, every person who has ever been treated as an outsider has in some ways been typical: representative of a community that they may never even have met. It's the smallness of our individual and social worlds that conceals this: persuading us we are seeing the entire system when in reality we only ever glimpse a tiny subset, drawing misleading conclusions as a result about average behaviour and 'normality'.

Within ergodic theory there are multiple branches of study that explore which systems do and don't fulfil the criteria. But the essential point is the important one to grasp: *any* large enough sample of people on a Tube carriage, crossing the road or putting down towels on the beach will ultimately be indicative of the average behaviour of other people within the same system at another point in time.

Consider that and then think about the individuals who make up your sample. There will be people of all shapes and sizes, races and genders, neurotypical and neurodivergent, with and without mental and physical health conditions. This slice of average contains all of us – in all our weird and wonderful diversity. You might call me crazy (plenty have) but I am as much a part of the indicative sample as you are. The whole system, moving in its consensus direction, contains all number of variances between individuals. Our differences remain strong and defining, even as we are essentially trying to do the same thing, and squeezing our divergent behaviours into an overall mean.

As a demonstration of human behaviour, the crowd is doubly ironic. From a distance, we see a homogeneous bloc and tend to overlook the individuality that facilitates the whole. While up close, within the heat and noise of the crowd itself, we see only the individuals, and lose sight of the collective movement they create. The assumptions we make as a result can easily end up backwards – seeing difference as a problem rather than a contributor, and assuming that consensus behaviour should trump individuality, when in fact it depends upon it.

Learning about ergodicity helped me to see that the human obsession with stereotypes is one of our more harmful traits. We rush to categorize people into distinct boxes to which we assign particular assumptions and expectations, often negative. And we then use those artificial categories to demonize people, emphasizing difference as a social and cultural weapon. Ergodic theory reminds us that there is a category, and we're all in it: the human race. It's within that capacious box that our similarities and differences should be considered – respecting the delicate balance of consensus and individuality which is the essence of being human. Any attempts to do otherwise are as disrespectful to the science as they are to people.

It is so easy to draw the wrong lessons, ones which enhance division and discrimination, and so important that we spread the right ones: an understanding that it is the sum of our individualities that makes us whole, and that an overall consensus depends on people breaking the rules as well as following them. We need people who deviate from the mean to explore ideas and places no one else has gone to. A mainstream will wither without its outliers to refresh, challenge and extend the overall consensus. Everyone has a part to play (even hipsters).

Embracing diversity in this way is something that has been essential to human survival through centuries of evolution. The same is true in our bodies, where cancer cells rely on their mutational outliers to accelerate progress: it's the side branches – subclones – that make cancer so hard to treat, because they allow it to adapt to different scenarios and respond dynamically to attacks. Cancer's diversity of structure is what gives it options – and the same is ultimately true of humanity. We rely on the outliers to evolve and avoid the stasis of the bystander effect, where everyone simply copies each other and no one goes to help the person in need.

Ergodicity is something that has been hugely important to me. As someone who grew up feeling like an island, it took me a long time to even glimpse the other coastlines, let alone build bridges to them. I have had to model the dynamics of the crowds in my life from the ground up – blind to the social nuances and 'isms' that instinctively inform most people's chosen paths. But when physics and probability showed me that even I – with all my weirdness – had to be part of the overall system, it allowed me to see myself in a different light. I know I am connected to the whole, part of the world's most powerful and beautiful system, the one that allows us to fulfil our evolutionary purpose as a species: to stay alive.

That has enabled me, as someone who isn't wired to connect with other people, to develop bonds of empathy with my friends and family. Because now I understand that all my extreme experiences – struggles with mental health, a sense of isolation and difference, prejudice from my peers – aren't a barrier between me and other people, but catalysts that allow me to connect better: a wormhole between different galaxies of living, mine and theirs. My empathy for people in difficult situations is an order of magnitude greater because of everything I have gone through, and the advice I can offer as a result. I know I have lived in all these situations, and that connects me to the people in my life who are having difficulty. I can quite literally envisage myself in their situation. Ask anyone who is neurodivergent or lives with a mental health condition: endless endurance and innate adaptability are our hallmarks. ASD and ADHD are my qualifications every bit as much as PhD.

Empathy has to be a balancing act, because if we give too much, we risk sacrificing our own endeavours on the altar of others' needs. Some people want to make you feel selfish when you are only trying to protect your own time and priorities. I might want to build a bridge from my island, but that doesn't mean I can cope with everyone crossing it whenever they feel like it. That said, since I started to experience empathy, it has become almost like a drug for me – something I didn't have access to for so long, and now pounce on at every opportunity, like someone who spent years without seeing light or tasting food. For years I yearned for human connection, to show that I am made from love, and that those like me – who are deemed by some to be crazy or abnormal – are actually some of the best, most non-judgemental people you will meet. I think of empathy as a painful euphoria, because it hurts like hell at times, but it is also something that no other feeling or experience can replicate.

So when the phone rings at 10.55 p.m., the time when I am usually in bed, putting my demons to sleep and going over my aspirations for the day ahead, I jump to answer it. I step out of my most peaceful place because I know a friend's world is falling apart and, as someone who has lived through the same experience multiple times, I can help her. And over the next two to three hours, I hear the light coming into her voice. Which for me is the greatest feeling in the world. Every painful experience I have ever had then becomes a valued treasure – a currency for empathy, something that connects me to other people who need what I've learned. The things I once considered to be irreconcilable differences with humanity now become my means of bridging from my island to theirs.

Ergodicity is the mathematical theory for anyone who has ever felt alone, different, isolated or abnormal. Statistics is telling you that your individuality matters, as much as anyone else's. It is a part of the weird and wonderful diversity that humanity relies on for its evolution and survival as a species. Quite literally, it counts.

Throughout our lives, individuality and conformity exert equal and sometimes opposite forces on us. The desire to stand out, and the need to belong, exist as parallel urges in all of us. We are individuals who can only survive and thrive in the collective context.

Everything I have studied about crowds over twenty years has led me to a clear conclusion. This is a duality we should embrace rather than seeking to fight. There is never any ultimate victor in the tussle to create equilibrium between me and we. Both have an essential role to play in our lives, and both must be respected. Both have something important to offer us.

What's more, neither is going away. Our individual personality and character will always be in us, however much we might try to alter it. At the same time, retreating into our-selves as individuals doesn't make the world go away. However much you might try to live on your own private island, there is no such thing as the entirely independent life. We have emotional and practical needs that can only be satisfied by tapping into the collective. At some point, even those of us who embrace our solitude have to leave our own shore, otherwise we never have anything to compare our solitary endeavours to. (And if you don't relish the departure, it's much more likely that you will enjoy the destination.)

As a child, this was something I feared above all other things. My mum used to say that going out with me was like a circus act, as I contorted myself to avoid the touching, and the sounds, noises and smells that scared me. But even though crowds still make me anxious and afraid, studying them has been one of my most important and beneficial experiments. It has helped me to recognize that individuality isn't every-thing, nor is it something that you should ever deny or feel ashamed of. I can remain myself and keep hold of my person-ality, at the same time as being part of a wider world that I both benefit from and contribute to. Participating in the col-lective doesn't stop me from being myself – in fact it makes the most of who I am, my experiences and what I have to offer. A dash of conformity has not detracted from my indi-viduality, but deepened it.

My attempt to analyse crowds was born out of a need to cope with large numbers of people. But in the process I learned that I can do more than survive among other people. I can also connect and offer something unique. And the same is true for all of us.

7.   How to achieve your goals
Quantum physics, network theory and goal setting

It was my first heartbreak. I was eight years old, and he was the thing I felt the closest connection to, other than my dad's fried noodles. You may remember him, as he was quite the name in science. I'm talking about Stephen Hawking.

It would be hard to overstate my childhood hero worship for the greatest physicist of my lifetime. From the way I ate to how I looked out of a window and sat in a chair, I tried to copy him. It even got to the point of me emulating him as my chosen hero in drama class. I told you, I was in deep.

But then my hero disappointed, confused and upset me. I was reading his most famous book, *A Brief History of Time*, specifically the second chapter which deals with space and time. Here, he explains how the historic belief of space and time as fixed entities has given way to an understanding that both are dynamic, shaping and being shaped by the objects that pass through them. Space and time are neither fixed, infinite nor independent of each other. To understand the universe we must visualize them together as four dimensions: the three of space, and one of time.

Hawking uses the image of a light cone to visualize this concept of 'spacetime', and demonstrate how past and future events are connected. When light is emitted it spreads out like ripples in a pond, forming the cone shapes. Because nothing

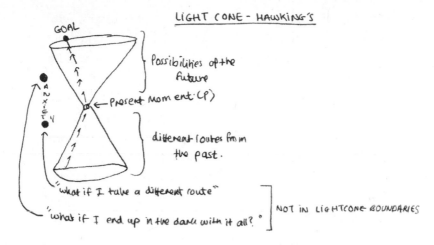

LIGHT CONE - HAWKING'S

GOAL

Possibilities of the future

← Present Moment (P)

NEXT x-EAT y

different routes from the past.

"what if I take a different route"

"what if I end up in the dark with it all?"

NOT IN LIGHTCONE BOUNDARIES

can travel faster than the speed of light, every event that contributes to (past), or stems from (future), the present moment must therefore be happening within these cones: at or within the speed of light.

Events that take place outside the cones are said to be elsewhere: therefore they cannot change the present, nor be changed by it. To illustrate, Hawking offers the scenario of the sun suddenly dying: this has not happened in the past light cone, and it does not affect the present because of the eight minutes it takes for light to reach us from the sun. Only at this point, some distance into the future light cone, does it intersect with and change our reality: we acknowledge the event not when it actually happens, but at the moment it starts to cross our consciousness.

Reading this for the first time, I did not feel my customary rush of excitement at a new concept I could explore and use. I was used to science illuminating my world and helping to explain it. Now I was faced with the cold, diagrammatic reality of a vision that clashed with mine. Here was the future as a fixed, quantifiable entity, sketched out in solid lines, while my vision was all about wobbly boundaries, interlinked outcomes and adaptable possibilities. This incongruence was like suddenly finding that your house key no longer fits in the door. Instead of feeling comforted and intrigued, I felt choked and my anxiety kicked in hard. It was as though my vision for the future had been whitewashed. What was happening beyond the boundaries of time in this model? What if I ended up out there, outside the cone and blinded beyond the light?

This was a scary moment but it was also a galvanizing one. It was when I realized that I couldn't get all the science I needed from other people's books and theories. To make the world make sense, I had to use my individual perspective. It was from this point that I started writing down notes in my own words, combining what I was learning with reality as I experienced it. I was oblivious to what the use of this might be, but it felt right and necessary. Now it's the book you are holding in your hands.

And I couldn't have started my journey with a more important topic than this. Thinking about how the past shapes us, how we experience the present, and how we can shape our future, is as fundamental as it gets.

We are all looking for ways to learn from what has happened to us, and influence what might happen next. We want certainty but also opportunity: to feel safe about our future but also to be inspired by the possibilities. While accepting that there are things we cannot influence, we want to know about the ones we can change. We want better ways to set goals, make judgement calls and fine-tune our priorities. We need a way to live in the moment as well as tools to plan effectively for the future.

The good news is that these aren't just questions we ponder while lying awake at night, or when we sit at the beginning of each year to write down our goals and resolutions. Theoretical physics has done a lot of the heavy lifting for us. It shows there are ways of visualizing the events in our lives that can help us to plot a path forward and maximize the possibility of a desired outcome. Even better – as I would have liked to reassure my eight-year-old self – these are methods that don't rely on the binary model and harsh boundaries of the light cones. By using the ideas I will introduce in this chapter – network theory, topology and gradient descent – we can all use methods to plan our lives and set goals that are as flexible and changeable as we are.

The big question: now or later?

When it comes to life planning and goal setting, perhaps the biggest question we face is what to focus on. Should our emphasis be on the present or the future? Is it about gratification now or pleasure deferred? Does constantly planning for the long term inhibit your ability to enjoy the here and now; or does too much focus on the present mean you will be ill prepared for what comes next?

Or, is it possible to have it all: a life well lived in the present along with one ideally planned for the future?

If you've ever worried that you struggle too much with this dilemma, quantum mechanics – the study of subatomic particles, the smallest we know about, and a subset of theoretical physics – is here to reassure you. Heisenberg's Uncertainty Principle tells us that the more precisely we can measure the position of one of these particles, the less effectively we can track its momentum. And the same applies in reverse. In other words, physics tells us that we can't measure location and speed of movement with accuracy at the same time. The more we focus on one, the less precision we can achieve with the other.

Sound familiar? Heisenberg may have been writing about quantum particles, but the same principle seems to apply at the macroscopic level of our everyday lives. Just as precision measurement equipment has its limitations, so too does our ability to concentrate and prioritize. You can't be a great host at the same time as enjoying the party: you're either thinking about it, or experiencing it, having a great time yourself or worrying about whether other people are. One detracts from the ability to do the other. Especially if, like me, you've had to prepare by googling, 'How to have fun.'

This is the dilemma of adulthood, where we're constantly aware of two contradictory needs: to live in the moment and to plan for the future. The desire to do both at once eats away at the capacity to properly achieve either. We're either dragging ourselves away from enjoyment by worrying about what comes next, or having such a great time that we never get around to organizing things for the future. Even as someone who relishes taking an information-driven, research-based approach to life, there are times when I just want to unlearn everything and be a kid again – basking in the bliss of being ignorant about the world, and the ability it engendered to truly live in the moment.

Alongside the diligent researcher, there lives a part of me that yearns to return to family holidays in Cornwall, the time in my life when I felt most liberatingly, unrestrictedly alive. Even the car journey to get there was an event. After three hours of driving, two packets of crisps and fifteen games of I Spy, we would finally reach the point of peak anticipation: when Devon gave way to Cornwall, marked by ecstatic screams from the back of the car as my dad drove us across the Tamar bridge. 'We are in Cornwall . . . NOW!' With the county boundary behind us, there was nothing in the way of a week of Cornish pasties, fishing in rock pools and trips to Padstow.

These are some of my happiest and most colourful memories, a time and place where I knew how to enjoy myself without reservation. Cooking fish dishes in the kitchen with my dad, playing outside in the garden, making as many sandcastles as the heart and mind desired, sitting on 'Millie's rock' on Looe beach in my awesome, multicoloured swimming costume. My penchant, aged seven, for gingham, re-enacting scenes from films using my mother's Blue Denmark crockery, and imagining my future with my heartthrob: none other, of course, than Stephen Hawking. The colour, taste and smell of every memory remains clear in my mind, twenty years later. It was a time of doing whatever I wanted, not even considering what other people might think. The good life.

This mixed bag of hobbies might have been random, and seemed shapeless, but all form part of the past light cone that has led me to this point: an accumulation of experiences that reinforce my interests, identity and individuality. They remind me of a time when the fear of missing out, and worries about what will happen next, simply didn't cross my mind.

As children, we see time as endless, even boring: to be filled with whatever fun, colourful and interesting things we can lay our eyes or hands on. In adulthood, time narrows into a currency: something to be measured and parcelled up, and jealously guarded. By the time I was doing my degree, there hardly seemed to be any room for relaxation. There were final exams to prepare for, and beyond those, application deadlines to meet and a future to plan. My life seemed to have become an endless to-do list, leaving me with little time or choice but to keep ticking off the next item. In this context, trying to find a present moment just to exist and enjoy myself felt almost sinful, even if I could have achieved it. In these months, I lived almost on autopilot, numbing myself to emotion and denying myself the exploration and enjoyment that my inner child craved. Daydreams about Cornwall's beaches were interrupted by a voice that told me to focus on my studies, and the next planned unit of time, including the ones I had allocated to relaxation. Try as I might to escape, the voice kept commanding me back to the medical science library, out of the rock pools and back to the air-conditioned, lithium-lit hallways.

In my efforts to get the balance right, I have taken inspiration from another subset of quantum mechanics – the study of how waves move through space and time. This presents a classic Heisenberg problem: you can either pinpoint the way a wave is moving, or its position at a certain moment in time: try to put your finger on both at once and you lose track. To get around the problem we create what are called wave packets, grouping and visualizing lots of different waves together so we can study their aggregate behaviour. A single wave is hard to pin down, where a 'packet' of several can be studied more effectively. Setting goals and making life plans are not so different: in isolation, it's hard to see if any one decision or goal is the right one. We need the entire 'packet' – the full

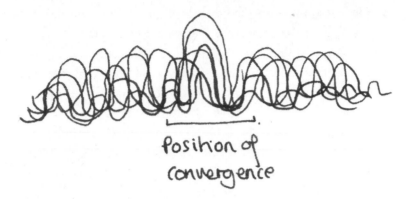

picture, and all the context – to understand if we are making the best possible choice, relative not just to the immediate moment but to our best impression of what the future as a whole looks like.

As we try to create these virtual wave packets, we have to strike another balance, between two different ways of thinking about our lives. There is *momentum* thinking, in which we live through time, going from one thing to the next, with our happiness defined by what we achieve and plan to do (the adult world of responsibility). And there is *position* thinking, in which we live for this time, captured by the present moment and the sensation it offers, blocking out everything else and simply being, including the feeling of guilt. This is hard, since it goes against the grain of what we are told it takes to be a 'functioning adult'. But it's also vital. Standing still doesn't mean you've stopped. Rather, it allows us to be more creative, reassessing our progress, living through the forces of our senses, and exploring more possibilities for the future.

Embracing position thinking gets harder as you get older, but it's still possible. My best moments are in yoga classes, where there is no noise, nothing to concentrate on except the posture you are trying to hold, and the opportunity to let all other thoughts and concerns dissipate, creating precious mind space. By the end of a class, as we are instructed to take on the shavasana (corpse) pose, I am too tired for any other thoughts to intrude. Often it leads to a cat nap on my yoga mat. Yet this rare moment of bliss doesn't come for free. Invariably, the next morning I feel sad: future thoughts and worries are again tugging my mind out of present harmony, even harder than before, sometimes to the point of self-punishment, be that via the withdrawal of food or cancelling social events to do something 'constructive'. I became a real a-hole – with myself as the main victim.

MOMENTUM → 1
LOCATION/POSITION

MOMENTUM THINKING

"what we plan for".

] single wavelength speed

I needed a way to break the cycle of momentum thinking, preoccupations of what's next, intruding on almost every aspect of life, denying me the joy of present moments. I wanted to recover my ability to live in the moment, without sacrificing my endless need for clarity about the future. So I tried an experiment, one inaugurated with a special batch of pancakes, just before Lent 2013: a socially acceptable moment to implement change. My own forty days and forty nights would be divided into two sections: half spent living in the world of momentum thinking, being completely rigorous about ticking every box and addressing every priority; and the second half basking in position thinking, trying to enjoy every single moment and not giving a thought to the future.

By now, you probably know me well enough to guess this didn't go particularly well (another of the all-important failed experiments that have made me the person I am). I couldn't avoid the thought of what I was missing out on – present enjoyment or future clarity – intruding on the experiment. I was hosting the party, but unable to stop thinking about the washing-up afterwards. I fell victim to another tenet of quantum mechanics, the observer effect: merely by observing a process, you inherently influence and change it – the classic example being that to observe an electron under a microscope, you rely on projections of photons that will change its course. Observing my own experiment, by definition, had skewed its outcome. I was too busy thinking about what I wasn't doing to enjoy what I was.

Since the failed experiment I have reached something of a compromise between position and momentum, present and future. At different times in a typical day, I will iteratively switch between the two, seeking to shift into whichever one I need the most at that particular moment. I battle my ADHD – which wants everything now, and has no concept of time – to try to manage the dance between living now and planning ahead. Just being aware of the Uncertainty

Principle is a helpful way of achieving the right balance. As I discovered, it's impossible to completely compartmentalize the two, but even the simple act of accepting their incompatibility is liberating. It helps us to worry less about the one we're not doing – realizing that there will be time for it later, and we shouldn't feel guilty about an afternoon in the sun (or one indoors planning, when everyone else is having fun).

But it's not enough to be conscious of how living in the moment differs from planning for the future, and trying to mesh the two mindsets. We also need a mechanism to visualize how present and future are connected, giving us clear choices about how to set goals and reassurance about the pace at which we are travelling. This is where network theory, one of my most trusted allies in life, comes into its own.

Network theory and topology

Ever since reading *A Brief History of Time*, I have been searching for a predictive model that suits my needs better than the fixed boundaries of the light cones. I was caught in the classic human contradiction – between the need for certainty and a frustration with setting limitations. Other than not knowing what happens next, nothing freaks me out more than actually having the limits of a plan imposed on me. I need flexibility: to turn those thick, straight lines into wobbly ones that I can navigate around and bend to my needs.

I needed a method of planning that understood both my need for endless preparation, to the point where it can take me five hours just to leave the house, and my tendency to tear up hours of careful thinking in a burst of intense impatience – a sort of psychological brain freeze, when the lemon sorbet you thought your day was going to be is turning into more of a vanilla ice cream. My Heisenberg struggle to reconcile present and future is sharpened by the time warp of my ADHD senses, with my mental accelerator constantly pressed to the floor.

In dealing with all this, network theory has been my salvation. This is a very straightforward concept: the study of how we represent connected objects via graphs, visualize the network they collectively create, and learn from what the connections tell us. It's what allows us to analyse complex, interrelated and dynamic systems by using the related techniques of graph theory.

A network is simply a series of objects or people connected together. You and your friends and neighbours are connected by a series of social networks. The London Underground is a network of stations connected by the different service lines. The electrical circuits in your toaster plug are networks. The smartphone sitting next to you is probably part of a network right now, connected to Wi-Fi and part of a WLAN (wireless local area network). The Internet itself is a mega network of computers connected both physically and wirelessly, moving vast volumes of data around.

Networks are everywhere, from the physical to the digital, the social to the scientific. They are the tangible and intangible structures that affect everything from how we build a career over decades to how we can get connected to the Internet right now.

They also provide the ideal mechanism for visualizing and mapping out our lives, over both the short and long term. We are all being affected by so many different things, pushed and pulled in a hundred different directions, that we need a more complex, iterative and adaptable model than a to-do list to plan ahead. Network theory provides this, especially when it comes to topology – how the different components (nodes) of a network are connected together, and the structure they form. Topology is what turns those inflexible, straight lines into a mobile network of possibilities: bringing some that had been shrouded in darkness back into light, and easing my anxiety of its pinch. It's what allows you to recognize that logic that may once have helped you no longer applies, or that an idea that has been germinating is now ready to flourish.

The nature of topology is crucial. If I gave you six buttons to arrange into a pattern, you could make a line out of them, or a circle, or a V-shape. This topology determines how the network will function: its capabilities and limitations. When we make decisions and set priorities in our lives, we are doing the same things: arranging the available evidence and choices into patterns that will determine both short- and long-term outcomes.

Thinking of our future life as one big network – with its nodes being everything from the people in it to our hopes, fears and goals – is the best method I have found for making plans that isn't either too simplistic or uncomfortably restrictive. It's helpful because it's dynamic, capable of adapting as your circumstances do. It's clarifying, helping us to understand what is and isn't truly important. And it's focused on connectivity – allowing us to see the things that are linked, which nodes are influencing or being influenced, and where a certain path may lead.

A network allows us to think as Hawking shows we must – in the context of both space and time – without being restricted by the tramlines of the light cones. It helps us to navigate proximity and distance – between people, specific goals, and stages of your life – across the dual canvas of space and time: what you need to happen, and when, and where you need to be for it to take place. Over time I have realized why the lines of Hawking's diagram exist, because we need directionality to create signal from noise, and overcome the anxiety of losing our way – becoming lost in our own life. But a network softens those lines into squiggles, turning the fixed cone into a leaf shape that can fold and curl up on itself as time evolves, exposing different sides of it to the light. It gives us structure, a path to follow, and also the flexibility to move around.

So the next time you are sitting down to write a plan, or worrying about what's going to happen, try replacing your to-do list with a network diagram. Treat every important person and goal as a node, and establish the connections between them: which people can help you to achieve which goals. Try to be realistic (in relative terms) about space in your drawing: which people or goals are most proximate, and which are distant? This is important, because you are looking for the junctures between different nodes as your path forward. The points where different components on your network come together are where you start to understand the unrecognized connections, and glimpse potential pathways ahead. You're looking for hubs – where lots of nodes are close to each other – and for potential elbows – where one path intersects with another, giving you a route. Also think about establishing an order of preference, for instance by colour-coding your objectives. A high-priority goal surrounded by lots of enabling nodes suddenly starts to look both desirable

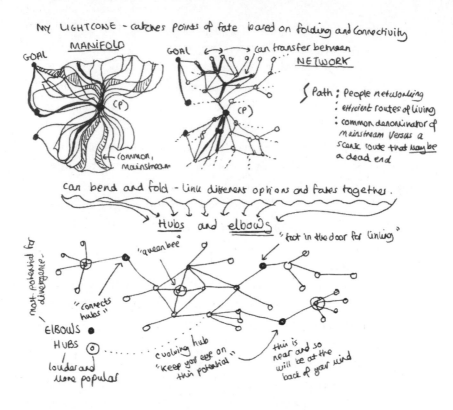

and achievable. In this way, the network begins to illustrate the things you want, the order of priority and what you can do to move closer to them.

Because – unless you are Stephen Hawking – it's incredibly difficult to think and draw in four dimensions, it's worth creating different networks for different points in time. One that shows where you are now. And two more that take you a few months, and then potentially a few years into the future. You might want separate ones for your professional and social networks too. This is something I do with my sister, Lydia, and we regularly help each other to set and refine our plans. We're ideal partners in this, because she is a perfectionist who is brilliant about pinpointing the immediate future, which I find scary, whereas I am good at looking into the more distant future, which she finds hard to fit into her closely controlled vision. I am able to help her think more flexibly about her long-term goals, and she helps reassure me about what is going to happen tomorrow – and what to wear that day. Because let's face it, I'd be happy wearing the same thing every day until instructed to burn it. We are both very happy and adept in our contrasting worlds, Liddy's where networking means the art of being well connected and meeting people, and mine where it means plotting nodes on a graph and establishing the probability of various outcomes.

In these conversations, something she has often said to me is, 'I want to do everything.' That fear of missing out is something many of us face. Surrounded by social media's hall of mirrors, we're more conscious than ever of the parties we haven't been invited to, the goals we haven't yet achieved, the 'gap yah' mountains we haven't climbed, and the feeling that our lives and our peer group are passing us by. What I always say in response is that she can do everything, but only by understanding how the different nodes are connected, and what should take priority, based on nothing other than what you want for yourself. It's simply not possible to do everything at once – but you can plan for how to achieve all the things you want. Over time the well-networked tortoise will outlast the frantic, fickle hare.

We need this ability to map across space and time – creating clarity about what needs to happen next – to avoid both overwhelming anxiety about the present, and gnawing fears over the future. A list of goals on its own doesn't help us, because there is no context, no sense of interconnection and no mechanism to establish preference. It can be good for the linearities of life, but to make decisions you need a network that plots your goals alongside people and places, something that doesn't have to adhere to any shape but your own. None of this, however, guarantees that we won't get anxious and envious as we overlap and compare our own topology with those of our friends and peers, wondering about all the would-haves and could-haves, and worrying about being left behind. Network theory can't save you from this feeling of FOMO, but at least it gives you a direction and a purpose, one you can shape flexibly and evolve over time.

Once you have your networks, you need to start navigating them: determining from the mass of information and components what represents a viable path forwards. How can you both identify and develop the optimum layout, and then continue to shuffle the moving parts as your situation evolves?

To answer that I want to look next at another machine-learning technique, and how it can help us to set a course from now to what happens next.

The gradient descent algorithm: finding your path(s)

Once you have your network sketched out, you can start to see the options in front of you. There are always different paths you can take, and you would be forgiven for being unsure about which is the quickest route to goal. Fortunately, machine learning is on hand to help. These questions of optimization – how to find the quickest and most efficient path – are the heart of computer science. Algorithms thrive on burrowing through data sets to discover how to do things more quickly, efficiently and cost-effectively. And we can borrow their techniques to optimize our own path through life – after all, they are based on human logic in the first place.

The algorithm used in machine learning to answer this question is called gradient descent. This is an approach used when trying to optimize a process and minimize its cost function (error). The analogy is of someone trying to climb down from a mountain into a valley. The goal is to get to the lowest point (minimum error) as quickly as possible. So the algorithm, which can't see all the paths at once, is programmed to navigate by gradient: continually finding the steepest downslope and reassessing with each step. As long as it keeps finding the path with the highest negative gradient overall, it's going to reach the bottom fastest. Just like people, algorithms of this kind vary in their attitude and approach. There are greedy ones, which choose the most rapid, immediate route – a bit like a politician trying to fit everything into a fixed term of government. And there are explorative ones, which persist with patience, allowing more routes and solutions to be tested out along the way. The latter is something I'm still trying to learn from – counterbalancing the greediness of ADHD to focus all attention on one thing while forgetting about everything else (which explains why I'm writing this in bed, in the middle of the night, still wearing my waterproof jacket).

Gradient descent is one of the most fundamental techniques in machine learning, and it's a concept with several lessons for us all as we navigate our own life networks. The first is that you won't be able to see the whole path, or even much of it, in advance. You can connect nodes and identify clusters, but ultimately our view gets hazier the further down the path – into the future – we look. And that's OK. Because the second lesson of gradient descent is that your immediate context tells you everything you need to know, right now. Just as the algorithm tests the gradient to determine its progress, we should judge the value of a particular path by our own metrics – is it making us happier, more fulfilled, more purposeful? We can't predict how something is going to work out in the future, but we can absolutely test the direction of travel, and go towards the one that minimizes our cost function in life – developing our sense of value and purpose, fulfilling the upper layers of Maslow's hierarchy of needs – which states that, once we have fulfilled our most basic human needs such as food and shelter, our focus shifts to more ephemeral matters, such as the ability to feel achievement, find respect, solve problems and be creative.

And if that direction starts to become less favourable – the gradient is tailing off and you have less momentum and feel stagnant, numb or just not quite right – then change it. A gradient descent algorithm isn't sentimental about its choices: it is happy to take two steps back if that means re-routing back onto the steepest path of descent. We should do the same. We need to be iterative in how we choose and adapt our pathways, changing course whenever we feel as though we are moving away from our goals and happiness rather than towards them. And we need to accept that there is no such thing as the one perfect path forwards: there's only the path we have the willingness, interest and patience to discover and then pursue. Your ultimate route is always going to depend on factors other than objective perfection: the time you have available to explore options, and how perfectionist you are as a person.

The gradient descent algorithm teaches us to identify a path experimentally, through trial and error, constantly assessing and responding to our environment, and not being afraid to retrace our steps. Its last important lesson concerns not the direction of those steps, but their length. This is a problem known as the learning rate. For the most accurate results, you program an algorithm to take the equivalent of tiny steps, inching its way forward and slowly accumulating findings. By contrast, a higher learning rate means you may reach the valley more quickly, but because the steps are less precise you may simply step over the lowest point. Fine-tuning the learning rate, so you get the best results as quickly as possible, is one of the biggest challenges of gradient descent. An especially difficult thing to do with ADHD, where time warps, context blurs and you end up making the most important life decisions while sitting on the loo.

There is no perfect answer to this, because it can change, just as there is no one optimum path through life. Everything is subjective, and you need to pick the right balance between speed and precision. The perfect path does not exist – in our life or anyone else's. From the available data, as represented in our network, there are numerous potential routes. As long as we let the evidence guide us, and keep hunting for the steepest gradient, we will find a way – in fact, I encourage you to have many ways. Just make sure they are ones which make you 'tick', and that you are equipped to do them.

Setting and pursuing our goals in life can be one of the most difficult things. There are so many considerations: should we pursue this ambition or that, optimize for the short or long term, do what makes us feel happy or what we think is most important? How do we create a vision for the future that is uniquely our own, and not beholden to those of other people? (One of the hardest things for a social, communicative species, but also one of the most important: living according to someone else's benchmark is a bit like eating with their spoon – it never tastes right.)

All this is enough to induce an anxiety attack, as I should know, having been through more than my fair share. And it's not just about the big, scary decisions in life. Last year I even failed to get my Mum a birthday card, as, having been to fifteen different shops, I couldn't decide which one she would like best. I got too anxious about the decision and ended up not getting one at all. This is explorative thinking, a testament to my love for her that inevitably left me empty-handed, and in the dark. Maybe I should have drawn the line at shop number seven?

But being anxious about the future – or 'not knowing' what to do next – can be a strength and not a weakness. Quantum physics and machine learning demonstrate that uncertainty, and a willingness to change course, are assets rather than liabilities. Not being sure about our progress in life is a simple facet of our innate inability to effectively measure momentum and position in parallel. While the willingness to change course is machine-learning best practice – where 'suck it and see' is key.

So if you worry that you haven't made enough progress in your life, or don't know what comes next, allow science to reassure you. Those fears are natural. And the anxiety is helpful, acting as a lens through which to simulate any number of different potential paths. I have always seen it as my supercomputer, allowing me to make links and see possibilities that others cannot. People have told me not to be silly, or that I'm off my trolley, but I wouldn't want to live without my anxiety and the ability it provides to scan the landscape, as well as the momentum it creates to learn more.

Setting and pursuing goals may be intimidating, but like any climbing challenge (a sport I love), it is just a matter of having the right equipment and personal endeavour. Heisenberg provides our belay, network theory our rope and gradient descent the route.

And remember, you're trying to climb down the mountain, not up it.

8.   How to have empathy with others
Evolution, probability and relationships

'Don't be so ridiculous, it's only an umbrella.'

Except it wasn't. To me, this solid little item wasn't something expendable, to be left behind in a café and replaced without a second thought. It was my security, my armour for the day ahead, its neat, hooked handle a comfort to hold on to in all weathers. The umbrella wasn't just to protect me from the rain: it could nudge away people who were getting too close and support me against stairway banisters I couldn't bring myself to touch. Wherever I went outside, it came too: a mascot and a guardian. It was as important to me as a flash car or heirloom watch might be to someone else. Because the concept of money, other than as a means to survive, is mostly lost on me, the things I value are the few possessions that I trust as dependable companions. My umbrella was perhaps the most important of these.

And now it was broken, and the boy I was dating wanted to tell me it was only a silly bit of nylon and wood. He was unconcerned; I felt like crying.

The breaking of the umbrella might have been the breaking point between us. It threatened to be the moment that occurs in every failed relationship, when it becomes clear that one partner doesn't respect or understand something that really matters to the other. As humans, too often we lack the empathy to see the world from someone else's perspective, and impose our own beliefs on them. The gap grows between the person we want and expect to be with, and the one we actually are.

I told you, it was much more than just a silly old umbrella. The boy in question quite quickly cottoned on to this, so ended up outlasting my beloved brolly. But I had been reminded, yet again, of the difficulty of sharing a life with someone when you live in fundamentally different worlds.

Relationships, whether romantic or otherwise, are something I have had to work hard to understand and navigate my way through. Living in my own head is difficult enough, without having to do the same in someone else's as I try to work out what they are thinking, what they mean and what they want. In fact, you might find it strange that I am talking about the importance of empathy, a subject on which us Aspies are supposed to be clueless. If there is one phrase you get sick of hearing, it is, 'Try to put yourself in their shoes.' The assumption is that, being autistic, we need all the help we can get to feel empathy and relate to other people.

But if I've learned one thing, it's that people who are good at talking about the need for empathy often aren't much good at showing it. Whereas, although I might not *understand* why someone thinks or behaves a certain way, you'd better believe I am watching closely and trying to figure it out. A lack of innate empathy means you have to work that much harder to divine people's intentions and expectations. Through my eyes, a relationship becomes a complex equation of trying to match my behaviour to someone else's anticipated needs. It's empathy by observation, calculation and experimentation.

That makes it sound simple, which it definitely isn't. Trying to understand, anticipate and respond to the whims of our fellow humans is one of the toughest jobs we have. The most serious detective work many of us will ever do is trying to establish what a hint of body language or an ambiguous phrase from a loved one actually means.

For this task, we need the best that science has to offer at sifting signals from noise, and deciding how to respond when the evidence is unclear. All relationships depend on an ability to read between the lines – to judge when it matters even if someone says it doesn't, or when something might not seem important but really is. To make these precise judgements, we need to fine-tune our understanding of evolutionary biology, acknowledging where our differences stem from, and how a relationship between people will evolve over time, just as our bodies did from a single stem cell. We need to harness probability theory to help us decide what is and isn't relevant evidence. And we can benefit from fuzzy logic (yes, that's the technical term), as a framework for judging a decision when there is no black-and-white, yes-or-no answer; and for managing the inevitable conflicts that crop up in any human relationship.

The empathy we need to build and sustain relationships is something we can find by looking at the fundamentals of how we develop as humans, and conversely, by adopting some of the techniques that have been designed to help machines function in a human world. For our relationships to prosper, we have to be not just at our most human, but at our most mechanical: able to calculate and consider as much as we feel and relate.

Getting started: cellular evolution

Both the strengths and weaknesses of our human relationships are based on difference. We are all shaped by our different genetics, diverse experiences and varying outlooks on life. Yet despite these numerous contrasts, we once all started as essentially the same thing: an embryonic stem cell, one that endlessly divided and divided to create the skin, organs, bones and blood that hold us together.

Stem cells are the ultimate in evolutionary wonder: single entities that can divide and specialize into any of the cells needed in the human body (multi- or pluripotent, if you want the fancy word). For example, all of the blood cells in our bodies have ultimately diverged from a common stem cell, via a process called haematopoiesis (one of my favourite words). This is something that's going on in your body right now, as we are topped up daily with the right balance of red blood cells to transport oxygen, and white blood cells to constantly update our immune system. It's their ability to split, re-form and renew that makes them the essential building blocks of humanity, and such an important part of medical treatments for blood and immune system disorders – helping to rebuild the body they created in the first place.

A stem cell is the foundation for every human, and it's also the ideal lens through which to better understand empathy in human relationships. Like a stem cell, every relationship essentially begins as a generic, unspecialized entity: two people seeing if they might like each other. Over time, where the stem cell divides into endless daughter cells with their own very specific uses, a relationship also becomes more defined and complex: an intricate web of shared experiences, understandings, language and unspoken meaning. Just like the stem cell, our relationships keep on specializing and differentiating over time – undergoing more mitosis (division) to meet newly encountered needs.

As we age, the repetition of this process starts to take its toll on our bodies. Every time a cell undergoes mitosis, it loses a little of what is called the telomere, the protective surface on the chromosome, capping the end of each DNA strand. In a process often likened to the gradual fraying of a shoelace, the telomere gets a little shorter with every division, until eventually it can no longer guard the DNA effectively, and the cell loses its ability to undergo mitosis and becomes senescent (inert). The tangible effects of human ageing, as our skin wrinkles and our organs begin to fail, is a function of this cellular withering. Over time, our cells

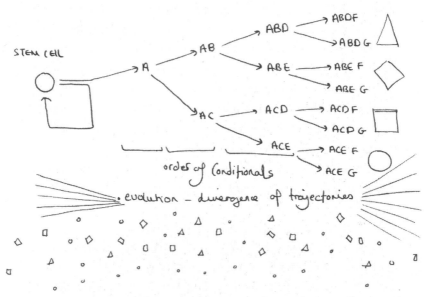

wear out and our body gradually loses the ability to repair itself.

Our relationships are subject to the same threat of decay, likely to die out if we lose the ability to undergo emotional mitosis, continuing to evolve and specialize in changing circumstances – as both our needs and those of our partner change. At the other extreme, a relationship can go too fast and become too intense to bear, in the same way that a cell which has mutated and can't stop dividing has become cancerous, growing out of control and starting to attack the body.

Understanding cellular evolution has helped me to realize two things fundamental to maintaining good relationships. The first is about respecting our differences. We might all look broadly similar and be part of one species, evolved from essentially the same ball of cells, but the devil is in the differentiation. Our endless evolution from that first embryonic stem cell has turned us into very different people. Often, it seems that the success of a relationship comes down to the ability of people to first recognize, then respect, these differences. It's through our empathy with other people that we are able to establish the most meaningful connections, showing that we have really understood the people we care about, listening to what they are telling us not just in words but through small everyday gestures and non-verbal indicators. This at times may even involve eye contact. (The things I do for human connection.)

Our closest relationships are those in which we can rely on these things in abundance – allowing us to feel known, appreciated and loved without reservation. So when my sister was getting married, and I had to pick an outfit for the big day, we both instinctively recognized the scale of the challenge. I knew how important it was to her, as someone who works in fashion, that the choice was a good one. And she knew how much I hate shopping, and that I was prevaricating not from a lack of care, but because I simply didn't know what to do. Being sisters we understand each other, so I was saved the ordeal of a solo shopping trip, and she was spared having her maid of honour turn up dressed, after Jim Carrey from *Dumb and Dumber*, in an orange tuxedo and top hat. ('Don't wear a top hat.' 'But you said I could wear anything I want, and you love Jim Carrey.')

The second lesson from cellular biology is about patience. Just as it takes the embryonic stem cell nine months to gestate, and then a newborn baby up to eighteen years to complete its physical development (and a few more, neurologically speaking), a mutual understanding in a relationship cannot fully mature overnight. If, on the second or third date, we are already starting to envisage the life we might share with this person, then we are imposing the expectations of a mature entity onto one that has barely started to develop. That creates an asymmetry between what we expect of a person, and what they can reasonably be expected to know about us. The safer bet is to treat a nascent relationship as the straightforward, hardly evolved stem cell that it is. Try not to project all your big, long-term expectations onto someone right at the beginning. It's all going to come crashing down if you ask too much too early of someone you haven't yet shared spoons with. Be understanding and have the patience to realize it takes time for the evolutionary process to bear fruit.

Probability and empathy: Bayes' theorem

The beginning of a relationship is really the easy part. As long as you don't let your expectations run ahead of reality, it's a time to enjoy the simple pleasures of something new and as yet unevolved.

But if you get beyond the honeymoon weeks and months, the realities of evolution must set in. As we get to know someone better, the single-celled organism of the first few dates starts to divide into something more complex. We acquire knowledge and shared experiences, and with them come expectations – that one person will know the other's mind, be able to respond to their whims, and anticipate their needs.

People sometimes talk about a relationship falling into the comfort zone, where partners stop paying proper attention to each other, but if anything the opposite seems to be true. If ignorance was bliss, then knowledge means responsibility. The demands on your empathy rise fast as the evidence you collect about each other starts to accumulate.

It's at this point, after we are meant to have got to know someone, that the real detective work begins. We are required to interpret small signals, half-hints and even complete silences. That's a potential nightmare for anyone, but especially when ambiguity isn't your strongest suit, and you are minded to take everything you are told completely literally. As an Aspie you have no preconditions or preconceptions when meeting someone: everyone gets seen with totally fresh eyes. So I needed a technique that could overcome my tendency to believe everything I was told, and my inability to naturally infer meaning from hints and signals. In this, Bayes' theorem has been my trusty ally. This is a branch of probability theory, concerning how we can use the evidence we gather to continuously evolve our estimates of how different situations might develop. In other words, as a situation changes, so too does your appraisal of the various probabilities.

As a Bayesian, your starting point also differs from classic statistical techniques. Rather than simply inferring probabilities from the data you collect – for example, the chances of a particular coin landing heads or tails, based on an experimental sample of it being tossed – you start with a series of prior assumptions. You use things you already know to help calculate the probability – which in the coin example might include the technique of the person flipping it, or the possibility that they are somehow trying to influence the result. Bayes' tells us not just to collect data and draw linear conclusions, but to put it into the wider context of everything we know about the situation in question.

Hold on a minute, I can hear you thinking. Isn't this the very opposite of what scientific research should be about: putting your finger on the scale instead of letting the evidence speak for itself? Well, it's certainly true that if your assumptions are wildly skewed, then so too will be your interpretation of the evidence. But there is also a simple but compelling power to the Bayesian approach: it lifts our sights beyond a narrow, time-limited data set, allowing us to widen our field of vision and put issues into a context that can otherwise be easily, and fatefully, ignored. For instance, it helps identify errors in medical screening: a test that may be 99 per cent accurate doesn't mean we have a 99 per cent chance of carrying a disease just because we have tested positive, but we only know this by using our prior knowledge about the prevalence of false positives.

Bayes' theorem allows us to consider *everything* we know about something or someone: when used properly, it is a brilliant technique for squaring the circle between what we know and what evidence tells us – exposing both potential flaws in our assumptions, and limitations in the data we collect. In other words, it helps the evidence to improve our assumptions, and the assumptions to improve how we use evidence. It's also important not only in how we approach questions of probability – making use of our prior, contextual knowledge – but also in how we update our assumptions as new evidence accrues. This is what's known as conditional probability – the chance of a certain outcome based on events that have happened, or still might.

Whenever I encounter anything new in my life, be that a relationship, a change of environment, or a new job, I use Bayes' theorem to help me navigate new uncertainties, and to tune into an unfamiliar culture and norms. I try to rid myself of my own biases and become a scouter, living based not on my own carefully curated preferences, but according to those that seem to be representative of this new system. When I went to university, I even put myself through the Aspergic nightmare of going clubbing: the deepest, darkest place no Millie Pang had ever been. I danced through it. And I collected data that was essential to giving me the new context I needed to interpret all the unfamiliar situations I was encountering, and the new experiences that were contained within them.

The same approach can help us when trying to find our feet

in a recent or evolving relationship. If we want to really under-stand someone, we need to observe them carefully, learning about the differences between what they say and what they mean, how they behave when they are happy or sad, what it means when they retreat into their shell (which could be sig-nalling a problem, or simply be an expression of their desire for space). That honeymoon period, when expectations are low? That's the time to be gathering all this evidence for when you will later need it. Someone will forgive you early on for not realizing that 'Sure, fine' means 'Definitely not', but over time that tolerance ebbs. The stage in a relationship at which we seem to have most latitude is actually the one where we need to be paying close attention – something that will pay off in the long run.

Of course the other inference of Bayes' theorem is that, because prior assumptions matter in how we interpret evi-dence, two people are likely to look at the same question differently. We need to have the empathy to understand that what seems like an open-and-shut case to us, could be the opposite for our partner – if they are starting from a different place of acquired knowledge, judgements and experience.

I also use Bayes' theorem to help manage the most tempestu-ous relationship in my life – the one with myself. However bitter an argument might get with a friend or partner, it's noth-ing compared to the raging tempest that is going on in my own head. Because my brain is having to work overtime to process all the data around it, considering everything from every pos-sible angle, it becomes a pressure cooker which can boil over without warning. Sometimes there's no alternative but to let out some of the noise that is pounding through my brain: bang-ing my head against the table, screaming and shaking, running around in circles. Anything to release some of the pressure of just trying to exist.

As well as routine to provide my anchor, Bayes' theorem has been my weapon in this very personal war. Rather than simply responding to the evidence in front of me – the noise, smell, or sight of a plastic button that would alone send me into meltdown – I can use my prior assumptions to drag myself back from the brink. That horrible smell can't really be so bad, because someone farted in class a week ago and I didn't die. The probability is, difficult as this might seem, that I'm going to be OK. Bayes' theorem has helped me to prioritize the different triggers that threaten my equilibrium, and separate the ones that are emotionally important from those which are just habitual pain points. It allows me to pick my ASD battles, and conserve some much-needed energy.

Human behaviour, whether our own or someone else's, is never wholly predictable and can't be quantified absolutely. But we can treat it as a question of probability, fine-tuning our knowledge and assumptions about the people in our lives, and using those to determine how we respond in different situations. Turning your partner into an object of scientific study might not sound sexy (to some), but it's the surest route to achieving empathy that I know. We have to do this because of the simple, and annoying, fact that people usually don't tell you what they actually want. They hint at it, signal it through body language or simply expect you to work it out. A nightmare when your mind, like mine, demands clear and unambiguous evidence to work from. Only by harnessing probability theory – applying what we know to the question of what someone *actually* wants, and what might happen next – can we truly find a way through these grey and misty parts of every relationship. So if you didn't know that an eighteenth-century Presbyterian minister (Bayes himself) was the best relationship counsellor you're going to meet, you do now.

Argument and compromise: fuzzy logic

Observing the people in our lives is one thing, but that only gets us halfway to solving the mystery of how we can meet both their needs and ours to build a healthy relationship, making the necessary compromises and overcoming the inevitable disagreements. We also need to be able to interpret the evidence we collect through observation, and take decisions that help to create equilibrium over time. This is where we can make use of one of the most important principles in artificial intelligence and computer programming: fuzzy logic.

You might assume that algorithms are a bad place to start when seeking techniques to navigate the grey areas in life. Isn't this the one field where the human mind is, and will remain, superior to the machine brain? Well, that would be the case if we were capable of using our maximum capabilities to discern complex situations, deploy empathy and make perfect relationship judgements every time. But unless that applies to you (and it certainly doesn't to me), it's worth taking a look at how the developers of machine learning have been grappling with just this problem. Perhaps an idea that's designed to help machines think more like humans can help us to do the same.

Fuzzy logic is the technique used to help an algorithm operate in situations where there is no certain truth, and not every factor can be categorized as either 0 or 1 – be that left or right, up or down, right or wrong. It allows a program to calculate in between the binaries, estimating the extent to which a non-absolute proposition is the case – for instance, whether something tastes nice or not. With fuzzy logic, an algorithm can determine whether something is mostly true or not, on a sliding scale between 0 and 1, rather than having to choose one definitive answer or the other. This has endless applications in developing automatic systems – from car braking that needs to determine how close the vehicle in front is, to washing machines that can adjust the flow and temperature of water, and the volume of detergent, during the cycle according to how dirty clothes are.

It also has applications in game theory and conflict reso-lution, as a methodology for mapping an ecosystem of different people with varying preferences, which may fluctuate between 0 and 1 – from absolute conviction to being totally willing to compromise – over the course of a negotiation.

It's this application that is most relevant for our personal relationships. However much you like or love someone, we all have arguments. The question isn't whether they will occur, but how ideally to manage them. Fuzzy logic holds the key, because it tells us that the human urge to 'win' an argu-ment is pretty useless. If something is worth arguing about, i.e. one person isn't willing to admit fault, then it's unlikely to exist at either the 0 or 1 end of the scale. Usually, it is in the grey area. Perhaps both of you need to apologize, or maybe there isn't a truly right answer about whether to get the new sofa in blue or red. An argument isn't a game, but more a problem to be solved, like a game of 3D *Tetris* in which you need to make the moving parts of your contrasting opinions fit as neatly together as possible. This is something I found because I've never been very good at 'fighting speech', and often didn't even understand the insults that other people would throw in my direction. That said, I can give it back in my own way: I'm no stranger to sharp opinions, and since I was described by someone at work as 'terse', it has become one of my favourite words.

We get into arguments for various reasons. Sometimes it's because we're bored – either in the moment, or of the entire relationship – and pick a fight to challenge and stimulate ourselves. But most of all, we aren't being bad actors or manipulators. We actually believe ourselves to be right, and our partner mistaken: a classic mismatch in intention and interpretation that is like clashing pieces on the *Tetris* screen. We

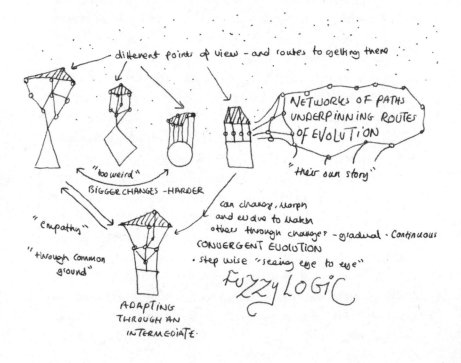

argue from a conviction that our assumptions and interpretations deserve to prevail. And we demonstrate a lack of Bayesian
empathy about what the other person may be feeling or thinking, or the accumulated experiences and assumptions that
have led us both to contrasting perspectives.

At this point, you can either engage in a shouting match
and door-slamming competition, or you can allow your
thinking to become fuzzier. You can accept that there is no
binary right and wrong in the issue at stake. And, like the
smart washing machine, you can adjust for the context. Maybe
this isn't something worth having a proper row about. Perhaps
you can alter what you thought was a strongly held view,
because it simply isn't worth it in the context of the relationship. Or you might just have to accept that you aren't going to
get what you wanted on this particular occasion, and it doesn't
matter as much as you thought it did. Conversely, there might
be some things that genuinely are significant to you, which
your partner doesn't immediately understand (aka my umbrella).
It's in these cases that you need to seek a compromise, rather
than offering one.

The most important thing is to lose neither your Bayesian
nor your fuzzy-logic perspective. Something isn't 'obvious'
unless two people are looking at it from an identical starting
point. Your belief that you are 100 per cent right about
something may fall apart the moment you realize that this
only holds when seen through the lens of your own unique
perspective, based on assumptions and experiences your
partner may not share. Having fuzzy arguments helps get
us away from the flashpoints of binary thinking and rash
statements – instead slowing down to consider all the options.
It's difficult to do during an argument, when emotion is at

the forefront, but if you actually want to reach a conclusion it's the better way.

Arguments can be a healthy part of any relationship. We all need the chance to air our feelings, much as a computer has to debug, identifying flaws and shortcomings so it can work more effectively in the future. An argument done well – respectfully, and fuzzily – can serve as a debugging process on issues that may have been inhibiting a relationship. It's an opportunity to show both empathy and vulnerability: revealing more of each other's emotional and personal tapestry, and understanding your evolution as people. But that can only happen if we learn what machines are now being taught – that when there is no absolute right or wrong, we need to find a way to live and make calculations in the grey space between. Understanding our biases, and being willing to flex our convictions in light of that self-knowledge, is crucial to helping any relationship over the hurdles of argument and disagreement. If you still love someone after you've had reason to hate them, then you have what I call an ideal relationship: something that only the unfolding of evolution can reveal. And the best part is, I'm as incapable of holding a grudge as I am at detecting irony or making assumptions about people. Five minutes after the argument has ended, I'll be in the next room offering you tea. Until next time.

At times in my life I have wondered if I am allergic to people, so strong has my negative response been to someone else's smell, touch or words. I physically flinch from behaviours I find threatening – which, you'll have worked out by this point, is a lot. I've often despaired at my ignorance towards my own species: my inability to relate to other people, or to feel part of their world. Much like my immune system itself, I am constantly updating my mental immunity to handle and embrace the evolving changes of people and life. Some changes are small and easy to fix and at other times it is a battle akin to curing the common cold.

But I also know, deep down, that it's love that makes us feel alive, even when it's inconvenient, painful and hard to bear. The mathematician in me is also a romantic. She believes there are ways we can use statistics, probability and machine-learning techniques to improve our search for love and harmony with the people we care about. And if you're sceptical about the role of data science in your love life, then I would ask if you've ever used Tinder, Bumble or any other dating app. Because the truth is that many of us have been sharing a bed with AI for some time.

Relationships may be far from a science, but there are many ways in which science can help us to manage them better. One is in understanding the vital importance of evolution – how it got us to this point, and how important it continues to be in all our lives. A relationship is never static and cannot be treated as such. It has to be respected as a dynamic entity, containing two (or more) people whose needs, wants and hopes are going to continue changing over time. Biologically, we are all hard-wired for evolution: it's what got humanity from cave dwelling to modern living, and transformed every one of us from a zygote in the womb to the adult human we are today. But we don't always understand or acknowledge evolution in our adult relationships. We behave as if people haven't changed over the course of several years, or as if it's inconceivable that they might. We don't always work to ensure that our expectations, assumptions and behaviours evolve to match how someone else's life is changing. So the first point is to be more conscious of evolution in relationships – our own and our partner's – and to respond accordingly.

The second is to accept the uncertainty and ambiguity fundamental to any relationship, and to find ways to work with it, rather than fighting against it. We can't simply demand of people that they be 100 per cent honest and clear with us all the time (much as I would really, *really* love this). We have to be smarter than that – closely observing a partner's behaviour to give ourselves the context and data we need to assess probabilities. Becoming better observers will make us better Bayesians – and ultimately more empathetic partners.

On top of being more evolution aware, and more probability savvy, we should be bias conscious: clear about how our opinions are shaped by our experiences, and how different two people's perspectives on the same issue can reasonably be. Fuzzy logic – an acceptance that the answer to most difficult questions lies not at either pole, but in between – is fundamental to achieving compromise and turning arguments into positive experience, not destructive ones.

We have all made mistakes in relationships, had regrets, and sometimes wondered what is wrong with us. We shouldn't beat ourselves up. Humans are complex enough beasts on their own, let alone trying to work together in a pair, or as part of a pack. But we can do better if we take a step back, and use some new lenses to look at the same old problems. Empathy, understanding, compromise – all these are things we are told we need to show in order to build lasting relationships. And all of them can be improved and enhanced through the kinds of techniques I have discussed. Believe me: if I can do it, anyone can.

9. How to connect with others

Chemical bonds, fundamental forces and human connection

Of all my school subjects, English was the one I always found hardest. At sixteen, I was categorized as having a reading age of five: not because I wasn't literate, but thanks to my over-literal interpretation of some questions in a comprehension test. (They asked what happened when a ball was kicked through the window; I wanted to know whether it had been open or closed.)

I would sit in a specially chosen spot that placed me as far from the teacher and as near to the door (and radiator) as possible. A five-star seat for a one-star class. My ADHD would spike as my mind pogoed from boredom to restlessness. As the latest passage from *Of Mice and Men* was read out, I would doodle my own version of the story, the only way I could understand the connections between characters and different parts of the narrative: a little language blending maths, art and literature that was all my own. My classmates, I could clearly tell, were mentally doodling as they zoned out from the reading. But it was my scribbles that caught the eye of my least favourite teacher.

'Camilla! It seems as though you are doodling again. Tell me, how would you describe the relationship between George and Lennie?'

'Tan(*x*), all the way.'

At this, quite a few woke up from their doodling and laughed. Emboldened, I continued.

'The tan(x) curve is one which demonstrates periods of great turbulence with short, calm plateaus in between; also harbouring a great contrasting symmetry, which at certain points contains regions that are polarizing, unapproachable and undefinable – asymptotes. I can see this about their brotherhood. It is almost magnetic.'

This, it quickly became clear, was not the desired response. I was castigated for not taking the book seriously, distracting the class, and even being a disgrace to literature (impressive, for someone who had barely read a novel at this point). As the teacher completed her rant, every head in the class now facing my way, she moved closer to the point where I could smell

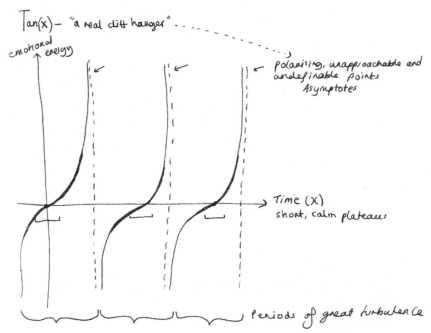

her breath. The tense silence and sharp odour triggered such an anxiety in me that I panicked, squatted under her armpit and sprinted through the door, hands over my ears.

But after the immediate panic had subsided, I started to feel triumphant. Out of the chaos of that moment, a new idea started to bubble to the surface, effervescent with possibility. My sketching, and unusual venture into literary criticism, had actually led me to something important. Thinking about the relationship as a trigonometric curve had sparked an epiphany. If a human (admittedly fictional) relationship could be represented like this, perhaps there were other ways maths and science could help me to understand the mysterious nature of human connections and relationships. These are the moments I live for, when I suddenly see a link between a scientific idea I know well and a human problem I have been struggling with.

The obvious place to start was bonds, the chemical attractions that link atoms and molecules together, and quite literally hold our world together. If people can be connected to each other culturally and emotionally, that is only possible because of the millions of microscopic chemical and electromagnetic bonds that hold our world and our bodies together. It is bonds that explain everything from the air you are breathing to the water in your glass. Without bonds, we and everything else would quite literally fall apart.

As well as being intrinsic, bonds are illustrative. Just as there are different types of relationships, there are also different types of bonds with varying properties. There are strong and weak bonds, temporary and permanent ones, some that are reliant on attraction and others that depend on a union of differences. What's more, like our human relationships, chemical bonds do not exist in isolation. Their existence and evolution are shaped by the fundamental forces that surround them, sucking them together, pulling them apart or moving them in different and unexpected directions.

What began as a classroom daydream has turned into one of my most important tools for understanding relationships of all kinds. Using bonds and force fields as my template allows me to model different relationships – explaining their shape, nature and purpose – and to understand the various directions they take, as you become closer or more distant with people over time. Crucially, it has allowed me to see that relationships are different. There are many kinds, with their own properties and features, which tell you something important about what expectations to have. Scientists rely on their knowledge of bonds to understand how different atoms, molecules and systems will respond to each other and react together. As people we could benefit by taking the same approach: understanding that, while no two relationships may be identical, there are broad categories which can tell us about the likelihood of different outcomes.

If we know more about the bonds that connect us with the people in our lives, and their distinctive properties, we are in a much better place to manage the evolution and growth (or demise) of our relationships over time. This one's for people who have wondered why a friend abandoned them, or who agonized over how to break up a relationship that had lived beyond its time. The answer lies not entirely in our actions and personality – or those of the other person – but in the nature of the bond that connects us. If we can understand that, things slowly start to make sense.

Introducing chemical bonds

All around us are chemical bonds, connections we can't see that allow everything we can see to function.

Bonding is the foundational activity of all chemistry. It's what allows atoms to join together, forming the molecules that – as we learned earlier – create structures such as proteins, the building blocks of the natural world.

Like in a human relationship, bonding involves the business of give and take – in this case, of electrons. These are one of the three kinds of subatomic particles that make up every single atom. In the nucleus (centre) of an atom you have positively charged protons, alongside neutrons, which carry no charge; while in the outer shells sit the negatively charged electrons. Together, these contrasting electrical charges mean an atom is constantly engaged in an internal tug of war, trying to achieve balance between competing forces – much as humans do in our own heads.

It's the exchange of electrons that defines the need for chemical bonding: to join with other atoms in such a way that will create a more stable overall structure: a compound. Other than the noble gases such as helium, very few atoms have the right number of electrons on their own to achieve peak stability. So they look for others to bond with, in ways that will complete them (aww).

In this, atoms are really no different from the people they ultimately create, looking for others to form connections with for a happier and, perhaps, easier life. And, just like our human relationships, the way they come together varies. Sometimes there is a true meeting of minds, in the form of an electron being shared; others happen when one atom gives up an electron for the sake of another; still more are the product of the electrical charges created as electrons are traded.

I believe there are clear parallels between the different kinds of bonds atoms form, and the relationships we create in our own lives. In understanding this, there are two main bond types to be aware of.

Covalent bonds

The most mutual form of chemical bonding is covalent, whereby two or more atoms share electrons in order to complete their outer structures. Within the outer atomic shells, the magic number is eight, the requisite number of electrons to achieve stability – a state in which the electromagnetic push and pull between the nucleus and the electrons is minimized.

Hence atoms are engaged in a sort of chemical speed date to find the right partner or partners to fill their quota. Take one compound we are all breathing in right now: carbon dioxide or CO_2. This comes about through a single carbon atom of four electrons sharing two electrons each with two oxygens, giving them both a stable eight.

Covalent bonding is an exercise in stability through sharing – a collaborative effort to create a chemical balance where both (or all) partners need each other equally. These bonds reflect the relationships in our lives that are based on common understanding and shared principles or values: where there is an innate symmetry that creates a long-lasting connection, and minimal drama or volatility. When you meet someone and feel like you've always known them, then you know how a covalent bond feels. The friendship is tight, immediate and reassuring.

Ionic bonds

Where covalency is about mutual dependency, ionic bonding relies more on give and take. Here, there is a transfer of electrons from one atom to another, creating an electrostatic charge that holds the atoms together.

In the case of another familiar everyday element, sodium chloride or NaCl, this happens when sodium donates the single electron in its outermost shell to chlorine, which has started with seven. Through this process, sodium becomes positively charged and chlorine negatively, and the two bond together through the attraction of their opposite charges. That's how you get table salt.

Ionic (or polar) bonds are those which are based on the attraction of difference. They are less about complementarity than the transfer of power. These are the relationships in which you know the other person may be totally different from you, but there is an interest or attraction that unerringly draws you closer to them. Ionic bonds are stronger than their covalent cousins, taking more energy to break apart – i.e. they have a higher melting or boiling point. This means that although an ionic relationship might be more emotionally volatile, in chemical terms ionic bonds are actually more stable. This natural asymmetry reflects the balance of power in a friendship, which in a healthy relationship equalizes over time through natural exchange and swapping.

The different kinds of bonds (these are the main ones, but there are further subcategories) show how the nature of the connection between us can govern so many different elements of a relationship: whether it is inherently strong or weak, formed from differences or similarities, and based on power sharing or a power imbalance. Like relationships, compounds

IONIC BONDS + COVALENT BONDS

→ "Handshake or holding hands"

→ "strong and stable" (but actually)

→ "push and pull is minimal"

→ "reliable"

→ "elements are equal in their desire for electrons"

"tunable"

"Polarity is life"

"natural push and pull"

"dynamic"

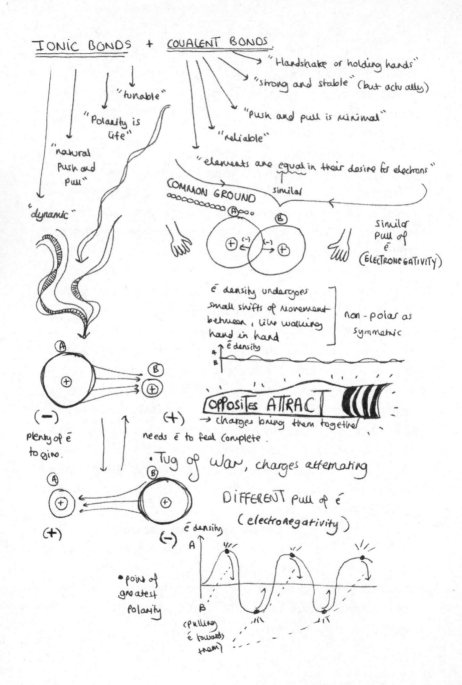

COMMON GROUND

similar

similar pull of ē
(ELECTRONEGATIVITY)

ē density undergoes small shifts of movement between, like walking hand in hand

non-polar as symmetric

ē density

OPPOSITES ATTRACT
→ charges bring them together

needs ē to feel complete.

plenty of ē to give.

(−)   (+)

(+)   (−)

• Tug of War, charges alternating

DIFFERENT pull of ē
(electronegativity)

ē density

A

B
(pulling ē towards them)

• point of greatest polarity

can be complicated and formed of different kinds of bonds together. The best example is water, whose core compound (H_2O) is the product of two hydrogens (1 electron each) bonding covalently with an oxygen (6). But it doesn't stop there, because the hydrogens continue to be attracted to neighbouring oxygen, forming additional ionic bonds – a combination known as hydrogen bonding. It's this mix of ionic and covalent that makes water one of the most versatile and accepting molecular mediators. Hydrogen bonds are akin to those you might have with work colleagues or teammates in sport: often not as strong as the connections with best friends or family, but essential bonds that can adapt to a wide variety of situations.

Just as protein personalities can help us to understand the different dynamics of a social group, understanding individual polarities is key to determining how people like to form relationships. There are extroverts who want to donate electrons, and introverts who would usually rather receive. And then there are the human equivalents of the noble gases – those whose electron shells (personal lives) are already complete, and have no need or desire for further interaction. Much as atoms go searching for one or multiple dance partners depending on their electronic need, different people may be searching for one partner to complete them, or for many friends to bond and connect with. This connective outreach, or atomic bonding potential, is referred to as their valency.

The hydrophobic effect

Conversely, there are people we can never get on with or to whom we become actively opposed – and bonds explain these antagonistic relationships as well. Think about when you add some drops of oil to water. Polar water comes into contact with non-polar, low-density oil. The result is that the two molecule types would rather interact with themselves than with each other. This is called the hydrophobic effect, and it explains why you should never drink water after eating spicy food. Instead of binding with non-polar capsaicin, the key compound of chillis, to wash it away, water simply flows past it – spreading it around your tongue where it binds onto even more of its receptors, heightening the burning sensation.

The hydrophobic effect also helped me understand the closed nature of some friendship cliques and the antagonistic attitude of bullies at school. People who won't let you into their circle, or who actively try to hurt you physically or emotionally, are defined by their unwillingness to bond. They have their atomically stable social structure, and want to continue interacting within that, keeping you on the outside. Human hydrophobia of this kind is about unstable atoms that bond together with others like them – defined by their non-polarity, the unwillingness to connect to others who might undermine the uneasy stability the group has created. What binds them together is a shared fear: of judgement, inferiority or rejection. Like the oil, they cluster together but remain isolated in the wider mix, fearing that to bond further with different people would pull apart the connections they have so far managed to form. As such, the social cliques that seem so desirable to our younger selves are often expressions not of strength and confidence, but weakness. They represent the fear of being shown up or outclassed by other atoms. Their defining feature is not what holds them together, but the mindset that compels them to stand apart.

As much as bonds provide us with a map for making connections, they also indicate where it isn't going to be possible – because you are dealing with fundamentally non-compatible molecules, who simply don't want to play together. If people aren't willing to come out of their atomic shell, there isn't much you can do. Bonds also reinforce the importance of balance in any relationship. Remember that every atom has a nucleus filled with positively charged protons, with its negatively charged electrons orbiting outside. When atoms get drawn together in an ionic bond, the two nuclei come closer together, with the distance between them being known as the bond length. The shorter that length, the stronger the bond. Unless, that is, the two nuclei get too close together in which case the protons start to repel each other. Just like when a friend or partner has become clingy, or dominating, or accidentally walks in on you having a number two – there comes a point when you need to re-establish boundaries. Bonds remind us that you can be both too distant and too close to someone to form an effective, stable relationship. It's all about finding the healthy middle, and understanding the inherent volatility of the bond you have forged with someone.

If you are looking to make friends, or find a partner, it's essential to understand both your own valency (combining power) and that of others. This helps you to judge what kind of connection you are likely to form, whether you are being asked to give, take or share a part of yourself, and what kind of relationship you have ultimately formed: on a spectrum from calm, mutual covalency to emotional, highly charged but harder-to-break polarity. It also lets you decide whether or not you feel open and stable enough to form that kind of relationship.

Ultimately, we connect with others for the same reasons that atoms bond: for stability, security and to obtain something that we lack when alone. But we can only form those bonds if, like the atom, we make them with the right partners and for the right reasons – and have enough innate stability ourselves to maintain them.

The four fundamental forces

Chemical bonds do not exist in a vacuum. They are the products of their environment, and the various forces in nature that hold them together and move them around. Forces help to explain why atoms come together, and how they fall apart. They illuminate how pressure is exerted and its effects over time. If we want to understand not just how a connection is initially formed, but how it lasts, or fails to, over time, then we need to understand forces and how they work. Of these, there are four literal forces of nature that are considered to be fundamental:

1. *Gravitational force:* This is the weakest force, but with infinite range. We all know how gravity works: it's the one that keeps us grounded. Without it, nothing would be anchored – you couldn't sit in your chair, the coffee wouldn't stay in your cup, nor the roof on your house. It's the constant, reassuring force in our lives: the reason I like to work sitting with my laptop

on the floor, where there is limited scope for anything to fall (as long as the ground beneath me holds). Because gravity is proportionate by mass – the greater the mass of two objects, the stronger the gravitational force between them. It's a force in which the largest object calls the shots. In the solar system, the Moon orbits Earth because it is only (just over) a quarter of its size. But the Sun weighs over 300,000 times as much as Earth, therefore pulling it into its gravitational force field.

We need to be aware of the same propensity in our relationships: are we equal partners with someone else, or does one carry a much greater mass of age or personality, and therefore act as the gravitational core? This might create an imbalance in which one person overwhelms and inhibits the other, or it could be that you are seeking just this kind of anchor in your life, and will therefore be attracted to someone who bears a reassuring mass (perhaps of experience) that you lack. Whatever the case, it's important to be aware that two people in a relationship are like objects imparting a gravitational force on each other, and to understand the equilibrium or otherwise that this has created. Usually there will be one person orbiting the other: it's helpful to understand which is which, and whether that is right for you.

2. *Electromagnetic force:* Helpfully enough for a chapter about relationships, electromagnetism is science's law of attraction, the force that brings objects together or drives them apart depending on the polarity of electrical charges. Electromagnetism, chemistry's

Romeo and Juliet, arises from two perpendicular
force fields. There is the electrostatic charge, where
two atoms quite literally have a moment – known as
a dipole – when their inherent polarity forges a
(hopefully not doomed) union. And there is the
magnetism part, where the revolving motion of
charged entities is so great it creates its own force
field, and what was a chaotic group of particles
becomes a coherent, directional entity with magnetic
potential. It's the power of these two forces working
together that underpins physical magnetism, the
natural world's fundamental law of attraction.

As we have seen, the electromagnetic forces created
by electron transfer are what create ionic or polar
bonds. Given that our bodies are hives of electrical
activity, it's no surprise that we experience much the
same at the macro level: an almost magnetic attraction
to some people with whom we feel compelled to
exchange electrons (and whatever else). These electromagnetic attractions can be stable or unstable, and
because of their polar nature, they can be polarizing:
excitable and highly charged. They are the essence of
our most exciting relationships, those where the
attraction is strong, some danger is present and there
is always the threat of volatility. So if you've ever
rolled your eyes at someone talking about 'the spark'
in a romantic relationship, cut them some slack.
They're being more scientific than they realize.

3. *Strong (nuclear) force:* As you've been reading this
   chapter, one contradiction may have occurred to
   you. If so much of chemical bonding is to do with

the attraction of oppositely charged particles and the repulsion of similarly charged, then what about protons? How do a bunch of positively charged, microscopic ping-pong balls hold together in the nucleus of every atom – doing literally the opposite of what their electrostatic charge compels them to? This is thanks to what's known, with winning simplicity, as the strong force. It isn't something you've heard of, but you're definitely grateful for it. Without it all our atoms would fall apart and so would we.

Without going into too much detail, and introducing new friends that sound like rejected *Doctor Who* characters (quarks, gluons and hadrons), it's sufficient to understand that the strong force both exists and is, well, much stronger than the electromagnetic one pushing the protons apart. So it's the most powerful, but also the most limited in range. In human terms, I like to think of the strong force as analogous to the most intrinsic, deep-rooted and powerful values that hold us together: love, loyalty, identity and trust. Like the strong force itself, we can hardly see these things and may not fully understand them; but we know how much they are needed as our anchors in life. As important as the bonds we form with others are, one of the most fundamental factors in any life is the strong force that comes from within, holding us together when it sometimes feels as though all the world is trying to pull us apart.

4. *Weak (nuclear) force:* The final of our fundamental forces is one that affects how particles change and explains much of the instability inherent in some

atoms. Not actually the weakest of the four fundamental forces (bad luck, gravity), it is also much
more influential than its name suggests. Operating
over a tiny range, it is the only force that can
actually change the internal composition of the
atom, promoting the decay of the nucleus. (For the
record, it does this by changing the flavour – or
type – of the quarks, which are the smallest
measurable subatomic particles in your protons,
neutrons and electrons.)

Weak force is responsible for atomic instability,
which is often the source of significant energy
releases. We need weak force to make the sun shine
(it causes the hydrogen nuclei to break down into
helium, generating massive thermonuclear force).
The same is true of nuclear fission. As that implies,
weak force can be the source of huge instability and
destruction. In our lives, it is sometimes represented
by the people who try to gaslight, guilt trip or
otherwise chip away at our confidence and sense
of self. Some people want to change you into a form
they will be more comfortable with – to bring you
to their level.

But in other ways the weak force is necessary, able
to break down a bond that no longer serves its
purpose, allowing us to move on from relationships
that have become difficult or toxic. Sometimes
breaking such ties is less an act of selfishness than one
of self-preservation. As volatile as change can be, it
also has the potential to unlock opportunity and
personal growth. In the same way as there are

fundamental forces holding and bonding us together, the weak force exists to pull us apart. You have to know when to resist, and when to accept it.

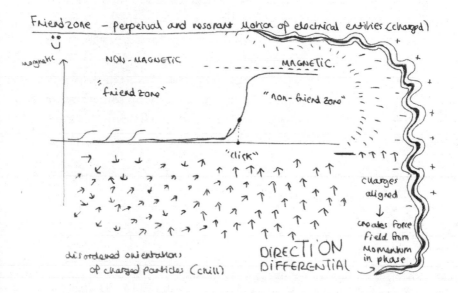

These four forces – the one that keeps us grounded, the one that creates attraction and repulsion, the one that holds us together and the one that enables things to fall apart – are fundamental to every element of our existence. And they provide a guide to thinking about how relationships come together, make us feel, and sometimes collapse without warning. They show us that it is the balance of different forces being exerted that is most important. When something doesn't feel right in a relationship, it will invariably be because an imbalance has been created: perhaps one person has lost their magnetism, or is providing too strong a gravitational anchor to allow the other to express themselves and evolve. Sometimes the strong force that held a relationship together may simply dissipate. Or one of you may have been compelled by weak force to change in a way that no longer makes you compatible.

When we're pondering what made us fall in or out of love with someone, or why a friendship that once really mattered has faltered, forces offer a good place to start. They allow us to consider more precisely what it was that brought us together with someone in the first place, and how and why those conditions may have changed. If the question is why something in your life came together or fell apart, then the four fundamental forces will usually hold the answer.

When things fall apart

If bonds provide a model for understanding how we as humans connect, they can also explain some of the reasons those connections fray and decompose over time.

No chemical bond is unbreakable. Every compound has its melting and boiling point, and the only real question is how much energy it takes. In the case of the ionic bonds holding sodium chloride together, a little water will do the trick, especially if it's hot.

Dissolving salt into your pasta cooking water may not sound very similar to a relationship breaking up or a friendship souring, but the essence is the same. The conditions in which the bond exist have changed and, as the temperature rises, the connection is no longer strong enough to hold together. All of our relationships will go through changes in circumstances: whether they are strong enough to survive will depend on both the nature of the bond, and the degree of change.

The hydrogen bond of a casual friendship is unlikely to last one person moving country, for instance. Whereas, if you have developed an ionic bond with a work colleague, it's unlikely that one of you moving jobs is going to stop you from being close friends. Your polarities as people haven't changed, even if the circumstances of your relationship have.

One of the most frequent reasons you hear for people drifting apart is that 'she/he changed'. We offer that simple, inadequate phrase to capture the full spectrum of personal evolution: how we change as we progress through life, enjoying success and enduring failure, and the imprint of our life experiences both good and bad.

Atomic compounds may provide a useful model to understand human connection, but of course we're quite a bit more complicated than that. Our needs, personalities and goals are liable to evolve over time in a way that the outer shell of a carbon atom isn't. It will always have its four electrons, and be looking for those two oxygens to complete itself with. As humans, our electrostatic needs are more fungible. We change, and with changes in personality, outlook and life ambition can come a change in valency. Looking for different things may mean looking for different people: perhaps steady friends in the place of party people, a partner who is focused on family as well as fun times.

I recently experienced what it is like for one of your most important friendships to break apart. This was someone I had known for years and shared the strongest kind of bond with. We'd sit around all day together, playing guitar and laughing until we almost peed ourselves. It was the easiest, most joyous friendship I have known. But our lives have diverged. Perhaps our careers have progressed at a different pace. The covalence that once so instinctively joined us faded, to be replaced by the sense that my friend needed something more from me, something I wasn't able to give. In a situation like this, it often feels as though the other person's weak force is taking over, denaturing some part of their personality or happiness, and threatening to take you down with it. You have to work out whether you are able – through sharing or donating electrons – to help that person re-complete themselves. But it isn't always possible. Sometimes either the magnitude or frequency of their electronic demand is too great, unsustainable for a healthy friendship. You shouldn't beat yourself up about it. Humans may be wired to connect, but there is a limit to how much we can offer other people without eroding the strong force that protects our own personality, needs and identity.

When you break up with a partner or close friend, the natural response (after having a good cry, obviously) is to blame yourself. You wonder what you did wrong and what you might have done differently. Bonds can help us reach a more balanced perspective, understanding that there are some evolutions no connection can withstand, and some bonds that were simply never meant to last, even if they played an essential role in your evolution to this point. Perhaps the most valuable thing is to know that seeing bonds break doesn't have to break us. In chemistry, by definition, a change in bonding or atomic identity is not just the end of one state, but the beginning of another: creating the space for new bonding potential. The same is true for us as humans. It might take a cup of warm milk to reset us and give us comfort after a relationship has broken down. But however many bonds we see come apart, we will always retain one of our most human abilities: to connect afresh, find new friends and love again. Our outer shell is ready to give, or share, its next electron.

The chemical bonds I have talked about here can form in a matter of nanoseconds, a time frame beyond our perception. Human connection can be pretty immediate itself, although we have to be mindful of the difference between these *affinities* (single interactions) and the biological concept of *avidity* – the connection created by the agglomeration of all those affinities over time. It is avidity that really connects people in a meaningful way, twisting two lives together through a web of shared experiences, interests, values and ambitions. Avidity of this kind only happens when two people can co-evolve, strengthening and deepening that initial bond so that we don't strain an initial covalence or magnetic attraction beyond its breaking point.

Nurturing these bonds is something we do instinctively. We will all spend time thinking about how to look after our friends, family and partner: finding the right words to support them at a difficult time, being there to celebrate their successes, even thinking what to cook for them or buy for their birthday. And, at the same time, we obsess over the arguments, mishaps and disagreements. Was it them or us?

Understanding our relationships through the lens of chemical bonds and fundamental forces allows us to see these questions in a new light. With this new perspective, we can better understand human connection, the factors that bring us together and those which force us apart. It helps us to comprehend the force we exert on others, and they on us – and whether that is a beneficial equilibrium or a harmful power imbalance. For me, it is about working out how to approach new relationships, and being able to reflect on those that have gone wrong for whatever reason, without instinctively resorting to blaming myself. Sometimes it's no one's fault. The bond breaks because of forces beyond our control. There is always that one tortellino that bursts in the boiling water.

Thinking about bonds allows us to reassess individual relationships and also to think in overall terms. These different kinds of connection nurture us in various ways: covalencies are about the steady, supportive relationships that give us comfort and reassurance; ionic bonds are where we find excitement, passion and often love. One is the river running steadily through our life, a bond that may ebb and flow, or change course, but which never runs dry. Another is the firework that lights up the night sky, thrilling us with its energy and possibility. We need both for different reasons, in a proportion that suits our personalities and life needs at any given time.

Just like the atoms that make us, we are constantly forming new connections, pursuing our fundamental human urges for belonging and stability. Some of those relationships will be fleeting, others lasting. Some will make us, and others will feel as though they are about to tear us apart. No one can ever be entirely dispassionate, objective, or dare I say scientific, about how they form new relationships. But chemistry can give us a new outlook and a fresh perspective: one that provides the confidence to make, break and sometimes re-make the connections that define us.

10.   How to learn from your mistakes

Deep learning, feedback loops and human memory

With ADHD, you're always forgetting what you're meant to be doing. My working memory – the part where we hold information for short-term, immediate use – is constantly undermined by new thoughts, impulses or emotional responses. It feels as though everywhere you go, even just to the room next door, the working memory is always being refreshed, losing immediate context. As well as making it almost impossible to hold a grudge, it also means I often leave the house to then forget where I'm meant to be going, or why, or to only realize at work that I've left my keys in my gym bag at home for the third time this month. I might get home and forget to take my jacket off for hours, because I suddenly got engrossed in a book I had picked up, or decided to erect some flat-pack furniture there and then. Or in certain cases I am focusing so intently on one difficult thing, such as planning my commute or working on a project, that I completely forget about anything else, like eating meals. My thoughts are like flies buzzing around the brain, loud but disparate, as opposed to the firmly anchored tent pegs of an organized mind.

Because short-term memory is a challenge for me, I've thought a lot about how the brain works to process and store memories. I've experimented on myself to see if I can improve the functioning of my short-term memory. And, as my understanding of machine learning has developed, I've begun to see how the artificial intelligence systems scientists are developing can help us to think afresh about our very human struggle with remembering.

This is important because memory isn't just about making sure you leave for work on time, with your keys, and wearing pants. Memory also represents the building blocks of who we are as people: the instincts, experiences and life events that have created the human being we are today, and will become in the future. Without understanding memory, we can't comprehend our thought processes, our psychology, our responses to people and situations, or what we value. In fact, we can't understand or fully know ourselves at all.

Conversely, a better understanding of how our memory works – what gets amplified and suppressed, what is near the surface or practically hidden from recall – can help us to achieve a more focused and supportive attitude to life. It allows us to escape the shackles of the bad memories that limit us, and focus on those from which we can learn or draw strength (cheesy, but true). Memory is something that can crush us if we let it: the accumulation of things we've done, said or thought that make us anxious or ashamed. These bad memories aren't just painful, they can actively prevent us from moving forward in life, such as the retrospective embarrassment of that time I got ridiculed for wearing blue eyeliner for a Tuesday lunch (a choice made out of pure boredom).

Much like energy, memories cannot be destroyed, only transformed (though unlike energy they can be created – as they are being in every living moment). Memory can take us back to the people and places that formed us, providing comfort and nourishment in difficult times – grounding us for our next venture.

Memory is intrinsic to all of us, and it's something we can take more conscious ownership over. It can be trained like a muscle – not necessarily to be stronger, but to better serve our needs, prioritizing the helpful over the harmful. We can become happier, more focused and more purposeful by developing greater awareness about how our memory works and aligning its capacity to our priorities. I learned how to do this by studying the workings of the closest scientific equivalent to the human brain – the artificial neural network. Just as these networks can be programmed to optimize how they process information to achieve certain outcomes, so we can fine-tune our brains to make more effective use of the oceans of data that our lives create.

If you've ever wondered how to escape the shadow of a bad experience in your life, to avoid the memories of what has gone before from limiting your future potential, then this chapter is for you. I want to show how, by applying techniques from deep learning and harnessing the power of feedback, we can leverage the power of human memory in our favour – learning from our mistakes without being constrained by our pasts. (Like that purple tank top I used to wear compulsively when I was eight.)

Memories might be made in the past, but their most important role is to inform decisions in the present and future. What we choose to remember is crucial to determining how we react to all kinds of situations in our lives. With the right, artificial intelligence-inspired tweaks, we can turn memory from a potential millstone into our most important source of power.

Deep learning and neural networks

Neural networks provide an ideal parallel to human memory for several reasons. The first, most obviously, is that they have been modelled on the brain: designed to produce the closest proxy to human intuition, perception and thought processes that artificial intelligence is currently capable of. The second is that their function is dependent on a feedback system that is crucial to understanding our own ability to retain and learn from particular memories. It is this feedback loop, and its implications for how we programme our own memories, that I want to focus on.

But let's start at the beginning. What is a neural network and what can it teach us about ourselves? Neural networks are algorithms that are programmed to turn inputs – senses and perceptions – into outputs – decisions and judgements. They are the principal tool of deep learning, a subset of machine learning that addresses complex problems which require the machine to 'think', working iteratively based on the data that is inputted. In other words, the algorithm uses the information or data provided to improve its understanding of a certain problem – which might be analysing traffic flows around a city, trying to work out how much house prices might rise based on historical information, or detecting someone's mood based on the look on their face. In all these cases, the ability to model good answers improves the more data you input into the system, and the more reference points the algorithm has to work with. Compared to traditional machine learning, a neural network is more independent and requires less input from the programmer to define what it should be searching

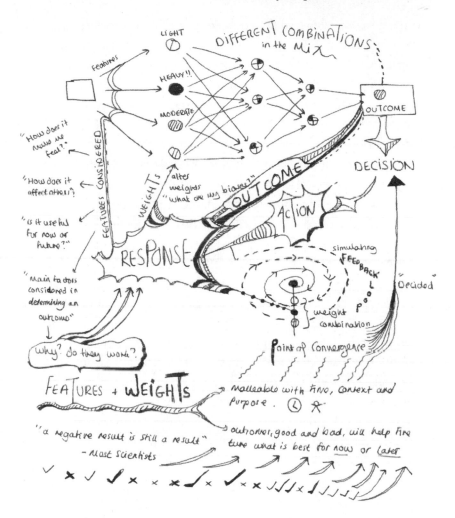

for, since, through internal layers of logic, it is able to create its own connections.

All of the more radical examples of artificial intelligence you may have read about – from fully driverless cars to mass automation of people's jobs – ultimately rely on deep learning, the closest we have so far got to developing a computer program that can think (within considerable limitations) like a human. Deep learning is also responsible for applications, including criminal checks, drug design, and the computer programs that rival the most competent chess players, all of which depend on an ability to simulate the connective capability of the human mind.

Modelled on the brain, a neural network is made up of neurons – in this case, the various data inputs. These come in three layers: input, output and in the middle what is called 'hidden', the place where the algorithm does its thinking. For example, if we take a driverless car, inputs will include the angle of the road, the speed of the car, proximity to other vehicles, the weight of passengers, and any obstacles on the road – all factors that determine the nature of the outputs, which are the decisions the algorithm makes about how to drive safely. It is the connections between these neurons, and how they fire, that really matter. The crucial aspect to a neural network is that the connections have a virtual 'weight' assigned to them, affecting their influence on the network and output. It's by comparing and calculating the weights of these inputs that the program reaches its decisions, learning which inputs to trust as most indicative of a particular result. With our driverless car example, it's likely to be speed and proximity of obstacles (which could be pedestrians or other vehicles) that have the heaviest weighting and the greatest influence on decisions. The ultimate aim of a neural network is, over time and extensive trial and error, to assign the most accurate values to these weighted connections, so it can consider new inputs with the priority (high or low) that they deserve.

So, rather than being able to tell the difference between a car and a pedestrian by feature extraction – isolating and locating wheels, legs, arms or wing mirrors – as a straightforward machine-learning program would, a neural network is able to use its weighted connections to simply detect which is which, and most importantly the combinations of data points that depict it most accurately (i.e., if it's got legs and arms, it probably isn't a Honda Civic). And the more images of cars and people you feed into it, the better opportunity it has, through trial and error, to optimize its weightings and combinations, and maximize the accuracy of its outputs (decisions). Much as we have different layers of memory piled up throughout our life, deepening our ability to establish connections and inform decisions, a neural network becomes more complex and sophisticated the more memories (data) it has processed. Like a child learning things for the first time, the more opportunity it has to exercise its 'mind', the better informed and evolved it will be.

This is thanks to its second crucial component: the feedback system. By comparing predicted and actual results, the network can calculate its estimated error, and then use our old friend gradient descent (turn to p. 139 for a reminder) to determine which of the weighted connections are most in error, and how they should be adjusted: a process called backpropagation (aka self-reflection). In other words, the neural network does something that humans are often bad at: it learns from its mistakes. In fact, it is hardwired to do so, without the emotional baggage that humans attach to their mistakes, using feedback as an intrinsic component of its quest to improve.

By contrast, humans often need reminding that feedback is important, and any engagement with it can be reluctant.

For many of us, feedback is a dirty word. In its most prevalent context, the workplace, it often acts as a neutral way to characterize a negative experience: being told that, for whatever reason, our work isn't good enough. It carries all the connotations of the awkward conversation, shuffled feet and words that don't quite mean what they say. But that's only because humans are, too often, bad at both giving and taking feedback. The neural network reminds us of its vital importance. Only by comparing what we expected with what actually happened, and adjusting our assumptions or approach as a result, can we ever get better at anything. If we rely on the same old weighted connections throughout our lives or careers, we will never change or evolve, or know why we are getting bored and frustrated by doing the same things in the same way.

When it comes to memory, we can all learn from the feedback-centric approach of the neural network. Or to be more precise, we can benefit from being conscious about this process, because it is one that happens already. The brain is busy weighting the information we process every single moment, deciding what we need to remember and whether it is for immediate, short-term retention, or something that we will always need to know. The things we remember are those we do or think about often (thanks to repetition), ones that are important (because then we actually stop and focus attention on them) or events and moments that have had a particular impact on us (also related to attention). Things that fall into these boxes get remembered – which isn't just to say that we store a memory of them, but that they themselves become part of the brain's algorithm, affecting our biases (weighted connections) and tilting the lens through which we process new information. What the brain considers to be important one day is going to continue to condition its priorities on the next. And vice versa. These connections and associations of memory are a tinted glass through which we view our entire lives.

To be unconscious about this process – the consistent ranking of everything we encounter into memorable and not memorable – is a bit like outsourcing your dating apps, using your previous preferences to determine automatically who to swipe right on, without actually checking whether you fancy them. It also leaves us open to errors in the system. In neural networks, errors can creep in if weightings have been based on too narrow a data set – also known as overfitting your model – or one that may suggest causation where only correlation exists: false alarms. So if you've trained a network to tell the difference between cats and dogs based only on paw size, it might be fooled by a really large cat or a small dog.

Our own brains are no less fallible. They may prioritize things we don't want or need to remember, or fail to log those which we do. We need the feedback loop to turn these 'errors' into data we can glean insight from, and use to make adjustments. As any scientist will tell you, there is no such thing as an error or a bad result, only sources of further learning. So if we want to re-programme our memory to better effect, we need to be more aware of the feedback loop that produces our core weightings, and start thinking about what we can do to optimize it. Without proper feedback, we are using only a fraction of our memory's capability to change how we see our life and the world around us.

Re-engineering the feedback loop

So what makes this feedback loop? How can we turn the light and darkness of our pasts into a memory that functions as friend and not enemy? We know that the feedback loop works: it's what has embedded past romantic haunts, that ugly cardigan we wore one time too many, and the most embarrassing events of our lives (you try letting your boyfriend use your iPad when you've forgotten to close a Google search for 'What are the pros and cons of getting engaged?'), into our consciousness. It's what makes us anticipate that first taste of coffee in the morning. But how do we make it work *for* us?

It starts with separating the mass of accumulated data from our lives into something useful. As humans, over time layers of memory have congealed in us, making it hard to separate different time frames and determine what is truly important here and now. Old bugs of the past can crawl their way into the present, clouding your judgement and ability to see clearly. Computers have the same problem – their memory getting clogged with too many programs running all at once. And they have a solution: to debug, getting rid of what is no longer useful or necessary.

To debug is hard for any human, but especially so for someone with ASD. While my working memory may be patchy, my recall for detail is the opposite – so effective that it gets in the way, as I become distracted from my present by remembering that a commuter on my train last month had a certain resemblance to an avocado. Being Aspergic means having the eyes of a hawk and the ears and nose of a bloodhound, neither of which is especially helpful at trying to be human.

Because we notice everything and store data from every detail of a situation, our memory logs soon start filling up. Letting go of this obsession with data collection is hard. Detailed memories are part of who we are: reaffirming our existence, and our connection to the people and places in our lives. And when you have a mind that is sensitive to every event and stimulus – not just when a car horn or ambulance siren is blaring, but when it isn't and you're waiting for the next one – it's not an option just to flick the off switch.

This is also a part of me I would never want to lose. The obsessiveness that manifests itself in endless preparation and routine is also the sentience that allows you to see the world in a different way: noticing beauty and difference where other people would never even take a second look. My ability to observe makes me open and alive, closer to my animal spirit than technological modernity often wants to allow.

But it's also challenging, because when you log every noise as a signal, it can be impossible to do as the neural network would, and establish a hierarchy of weighted connections. (It also makes for somewhat high-maintenance shopping trips: sorry, Mum.)

And, like everyone else, I have always wanted to fit in. I may have felt like I landed on the wrong planet, but that doesn't mean I want to live as an alien among natives. From growing up in Wales, to school in the Cotswolds, university in Bristol and jobs in London, I have worked diligently to try to swim in the mainstream. And one thing that I could never get out of my mind was the distinctively British reserve: the way people often talk and behave reticently, not quite saying what they're thinking, or trying to ignore something outrageous in the hope that it will go away.

I am not a reticent person. I am an enthusiastic giggler, a squealer in delight and a nostril-flaring howler in rage. My emotions have very rarely been a closed book to anyone. But I wanted to experiment. I wanted to become more reserved, to be more neutral and to enjoy the benefits of this seemingly more objective stance on life, which seemed to be the hypothetical optimum. Becoming less Millie and more English seemed like the opportunity to kill two birds with one scientifically aimed stone: a means to fit in, and a method to prune the tendrils of my overgrown memory bank. I wanted to be more like Siri or Alexa: all the knowledge, none of the emotional baggage. And people actually listen to them.

So I set out on an experiment. I would re-engineer my neural feedback loop, blocking off the impulses that make me the 'emotional weirdo' I am, and gaining the calm perspective of a completely neutral, very British being. This wasn't just shutting down my mental computer, but restoring it to factory settings: unlearning all of the associations that tangled around in my head, often constricting the clear thinking and the connection that I craved. In my mind, I could turn myself from an ADHD-wired, emotionally driven tempest into a gentle breeze of logical thinking and measured behaviour, who wouldn't forget her keys every other day, or have her views discounted for being too emotional. Through doing this, I would be able to disregard the unhelpful accumulation of old memories, and make logical judgements based only on the new inputs at that point in time. I would erase my biases, reset my neural weightings and start completely from scratch. A holiday in my own head.

But in trying to make myself remember better, I ended up forgetting things that were far more important. Over the course of this experiment, on a first date, a boy asked me what I was passionate about. And I realized I had absolutely nothing to say. I had made such a conscious effort to erase my biases and preferences, to get rid of those messy vulnerabilities, that I no longer knew what I cared about. I felt as though my spirit had fossilized, and I hadn't so much forgotten a few things as completely lost myself in the mental mist. I felt instantly, crushingly sad. And then scared. What had I done to myself? Ironically, by this stage, I couldn't even remember *why* I had set out on this course in the first place, since I forgot to make a note on my whiteboard. (Great, nice one, Millie. Once again, full marks for OCD consistency.)

As many experiments must be, this one was an almost total failure. It was a dangerous brush with trying to erase and deny my natural biases and true self. But, as failed experiments also tend to do, it taught me some important things. First, we have one spirit and personality, which is totally our own and should never be a source of shame or regret. We have to nurture that person, not deny or reject them. But, at the same time, we are not their hostage. I have learned to love myself as the unashamedly ASD, ADHD, GAD person I am: the full Millie. It has been the work of my life to balance out these competing parts of me, and capitalize on them where they are most useful. It's a full-time job in itself, a science and an art.

But that doesn't mean I don't find a lot of her behaviours a complete pain in the arse. The forgetfulness. The fear. The struggle to cope with big feelings. You can love a person while simultaneously hating being them. But, even better, you can chip away at the behaviour you find problematic. I forget things because my attention is constantly being pulled this way and that, eroding my instantaneous recall. I am afraid of smoke and loud noises because, deep in *my* neural network, the connections are irreversibly weighted by twenty-six years of having responded to these things in those ways. So my mental computer spits out a response to be afraid, and run away. These are responses conditioned by memory – its accumulation and its inconsistency. Which means they are problems that can also be addressed by training the memory and targeting the feedback loop.

I can't magically undo my forgetfulness or wipe away my fears. But I can find ways of managing them better, preparing myself for situations I know I will find difficult, and re-plumbing those neural connections to counterbalance the existing weights. It's a painful but rewarding and enlightening process: the human luxury of being able to fine-tune our own mental computers.

Some of these adjustments are very practical. Although my room might look disordered, it is actually full of clues to guide me through the day – starting with my dressing gown and toothbrush sitting by the right side of my bed, to remind me to get up, go to the bathroom and brush my teeth first thing in the morning. Others will strike you as plain odd. To remind myself to take any medications, I have to make an event out of it – shouting, 'Hagrid!' and doing a dance to myself. This routine might sound deranged, but at least it's memorable – adding weight to the likelihood that I will remember to do something really important, but too easily forgotten. And it's supported by a litany of Post-it Notes reminding me to pick up my socks, call my mum (twice) and not to wash the jeans that have £5 in the pocket.

Remembering to remember things is largely a question of finding the right mechanisms to remind yourself. Forgetting to be afraid is more complex. But this is about the feedback loop and backpropagation as well. Because I know that smoke or bad smells won't actually do me any harm, I can use that proven outcome to counterbalance the weighted connection that tells me to be afraid. I can try to update the inputs that condition how I respond to particular situations, by reassuring myself about a track record of outputs. This is never going to magically turn a negative feeling into a positive one, but it can reduce the intensity of the feeling, shift the dial slightly on that connection and enable me to retreat more often than not from the precipice of a panic attack.

You probably have quirks and kinks in your own memory and feedback loop: the past experiences that loom larger than they should (like a bad break-up), or the positive affirmations that we can over-interpret (just because you ultimately lived to tell the tale doesn't mean that last drink was a good idea). The important thing is to take some conscious ownership over a process that otherwise hums away unconsciously, robbing us of complete ownership of how we think about life situations and make decisions. If you are struggling to commit to a relationship because your last one was difficult, you need to remember that your previous one does not define you; the weighting of that may be too heavy in your feedback loop, inhibiting your ability to judge the new relationship on its merits. We need to think about *why* we feel a certain way, whether that is uncertain or over-confident, and try to locate the root of that emotion in the previous experiences that have filled our memory banks and conditioned our feedback loops. Once we have done that, it becomes easier to put both good and bad memories in their proper context and adjust the weightings accordingly: learning from our mistakes, getting over our hang-ups and looking forward to the future with something as close to objectivity as a human can ever achieve.

If we want to change how we feel about things, or approach particular situations in life, then the feedback loop is the place to start. We should recognize that our instinctive responses have been conditioned by a lifetime of memory and experience, creating the weighted connections that determine how our brain makes calculations. The things we value in life and feel strongly about haven't emerged by accident. They are rooted in our living memories, and the only way to change is by gradual adjustment via the feedback loop.

There are two kinds of feedback loop, positive and negative, and both have important roles in training systems. A positive feedback loop is an incitement to do more of something, giving it greater weighting and prominence in the overall calculation. These are for when we want to encourage a system (aka ourselves) to be bolder about something. Its negative counterpart is designed for the opposite effect, to restrict or limit a certain factor. Both have their benefits, and drawbacks. A positive feedback loop is stimulating, but can allow the inspiration and joy of living to spiral out of control, especially when it comes to drugs and alcohol, as we seek to revisit the same high our memory associates with them. Whereas a negative one, while acting as a stabilizing force, can also leave you tunnelling into a rut of introspection and futility: my experiences of depression have been when bad memories and experiences suppressed my positive energy to such an extent that I felt totally futile and unable to function for days at a time. This is the ultimate manifestation of a negative loop – where you experience an effective eclipse of every good memory and feeling that you have ever experienced.

If we are trying to create a positive feedback loop, then small doses of experiencing what we fear can build confidence, chipping away at the weight that tells us to hang back and be afraid. And we really can do things that make us afraid. I even once forced myself to go with my friends to a music festival (allegedly 'the best thing in the world'), which is basically the Asperger's equivalent of Mordor: excessive noise, endless mess, dubious smells and unpredictable crowds. It was there that I broke my personal record, of having five fully blown panic attacks in thirteen hours, not to mention the one when I accidentally got caught at the front, sardined in a mosh pit. After fainting from shock, I was crowd-surfed over to the medical tent for attention where they rang my parents. I then had to be rescued by my dad – who laughingly reminded me that us Pangs have never been happy campers.

I have used that experience to draw comfort from my ability to experiment, test my boundaries and remind myself that the unfamiliar (and even the unpleasant) isn't necessarily going to be fatal. You'll never catch me going back to Beach Break festival, or probably ever sleeping in a tent again, but I don't regret my abbreviated, abridged adventure for a second. I lived it up, however briefly, and I'd be willing to try something new like that again.

At other times we want to create a negative feedback loop, to stop ourselves from doing something. To achieve this, we might focus harder on the problematic outcome of a certain behaviour, reminding the brain of the asymmetry between the reason we do a certain thing and the end point that it invariably leads us to – be that a hangover, a sugar headache or being sick because you pushed yourself too hard at the gym.

These positive and negative feedback loops are constantly

pinging across our brains, whether we pay any heed to them or not. My experience is that, the more conscious I am of their existence, and the harder I work to re-engineer them (reminding myself of those good or bad outputs, contra to expectations), the more in control of my state of mind I become. It's also worth remembering that a well-functioning system, whether human or algorithmic, relies on the right balance of positive and negative feedback. We need enough positive to allow us to experience new things and learn, and sufficient negative to limit us from making silly decisions or putting ourselves in danger. We can't overdose on either positive or negative feedback if we want to maintain a form of equilibrium; just as the neural network powering a driverless car mustn't interpret information either too aggressively or too cautiously to be a safe driver. But neither can we do without them – both need to be deployed, and tweaked, to match the different situations we will encounter.

Through my experiment I learned that it's not possible to simply abandon a lifetime's accumulation of memory and mental preconditioning. Like it or not, these are the things that make us human, allowing us to feel and giving us an identity or personality that provides our anchor. These biases might sometimes feel like an enemy, but in reality they are just us – the purest expression of ourselves. But accepting the existence of these biases is not the same thing as surrendering to them. We can remain in control by being cognizant of them, working gradually to use real experiences to prime the feedback loop, and adjust those all-important weightings of experience. We have to bring our subconscious biases into the realm of the conscious so we know what we are dealing with. Like going through old photos, this is a process that can be both scary and hilarious.

Experimenting with my memory has taught me that wiping it entirely clean – and having to re-learn how to hold my favourite mug – is not the way to go. We can learn from neural networks, but we're not like computers that can make their memory work more effectively by getting rid of everything they have accumulated. In the place of a total memory format, I have settled on a continuous process of mental upgrading. Every one or two years I will look at the different layers of memory that are foremost in my mind – putting aside those which used to be useful but have now served their purpose, and trying to knit together those that give me inspiration, focus and happiness. It's a way to have fewer regrets about what is past, and to sharpen your mind for the challenges that lie ahead. At any one time, our memory represents the tapestry of our life. We shouldn't forget that we get to choose what it features.

We can't control everything that happens in our lives, but we can condition how the memory both stores and uses those experiences. Something entirely in our control is what we give weight to, how we remember it and for what reasons. What things in your life give you strength and remind you of the person you really are and what you are capable of? Conversely, what low points can act as speed bumps on future behaviour and decisions, reminding us of what we are likely to regret later? We get to choose and prioritize: the good versus the bad, the logical versus the emotional, our feelings versus those of others. All these elements are fizzing across the feedback loop on a constant basis: like any recipe, it is the proportions that matter, and we choose those by deciding what to really focus on, and how to process and store those memories. While learning from your mistakes might sound trite and simplistic, it is actually an important part of how we condition the mind and memory to work in our favour, creating a healthy, balanced feedback loop that better equips us to tackle future challenges – and not being afraid to fine-tune it over time.

The feedback loop happens instinctively, but there is a great power in actually owning it: making ourselves think about how it works and what adjustments we can make. Focus and attention are such important parts of how we form memory. Because my dad's lasagne is my favourite food in the whole world, when I was young I would always let it sit in front of me, taking in the smell and sight of it for ten seconds without blinking, logging the memory as one to come back to whenever I needed to feel the comfort of home. In small ways like this, we can all train our memory to prioritize the useful over the unhelpful, making the most of its capacity to provide strength, reassurance and comfort.

Memory can be a place of anxiety, shame and regret. Left to grow and evolve unconsciously, it encourages us to dwell often negatively on our experiences and past decisions, reliving them in the seconds, months and even years afterwards. Our challenge with memory is not getting stuck in its maze of history, regrets or people and places we can never return to. If we're being honest, most of us probably live more in the negative feedback loop than the positive: selectively accumulating the bad experiences and memories that chip away at our confidence and influence the trajectory of future judgements.

But memory is also something we can't live without. Something which, as I discovered, is too intrinsic an element of our humanity simply to strip out like a defective part in an engine. To format our memory, as we would that of a computer, is to wipe away too much that can never be properly replaced. So our best option is to fine-tune, making the adjustments over time that allow us to get the most out of this powerful, sometimes dangerous, source of ourselves.

11. How to be polite

Game theory, complex systems and etiquette

'Hi, Millie, I'm calling to see if your mum is there?'
  'Yes, she is,' I said, and then put the phone down. Job done.
'Was that the phone, Millie?'
  'It's OK, I took care of it.'
  Not quite 'getting it' has been one of the recurring features of my life. As an Aspie in a world of neurotypical half-meanings, ambiguous gestures and untranslatable implications, you are constantly walking through a minefield. Or you're accidentally sowing mines in their cornfield. Either way, it's a hazardous existence if embarrassment is something you fear.
  This never used to be a problem for me. I happily sailed through, treating everyone (and I mean *everyone*) the same. I said what I saw, told people what I thought and shouted at rude people in the street. Age, seniority and reputation didn't occur to me. But nothing comes without a cost. My carefree, context-free existence was one that didn't allow me to feel empathy for people and their individual needs. To do that, I developed the Bayesian approach to manufacturing empathy that I have described in Chapter 8. But in doing so, I lost the armour that had once protected me from feeling embarrassed. I became newly sensitive to human judgement, and to the demands of social etiquette. And then I started to

realize how big that minefield really was. Let me give you a few examples.

Of course, the first has to be dating. Now, I'm going to put my cards on the table. I can't flirt to save my life. But against all the odds, and those produced by the apps, I do date. One of these was with a person whose second name happened to be a pork-based product. He seemed nervous at our lunch so, in an attempt to make him feel at home, I decided to order salami. It turned out he was a vegetarian, and thought I was taking the piss.

Or there was the one time a guest at home asked for something 'a bit stronger' than tea, so I offered coffee. The incident in the train station when I thought a man had a bomb up his jumper and called the police, who made him turn up his top, to reveal that the deadly weapon was actually a hairy beer belly. The time when a friend of my mum's, who uses a walking stick, leaned in to hug me and I recoiled so violently that he ended up on the floor, and I had to flee the scene. And then later, when I tried to redeem myself by dancing to the Christmas music, I ended up poking one of my uncles in the eye with an overenthusiastic hand gesture.

Not to mention the occasion when I had a dinner date with my friend, who is like another sister to me. She is a fellow yogi, into her fruit, and I wanted to find the perfect gift to express my love and our connection. I duly arrived bearing a full-sized watermelon (an important part of the Ayurvedic diet). Cue confusion, and a later admission that it's one of the fruits she doesn't actually like.

As I've become more conscious of what other people think of me, I've also become more aware of the endless complexity of expected behaviour, and how that varies from place to place, group to group and person to person. Why does a joke that might be hilarious in one context cause embarrassment in another? Why is it accepted to eat a certain way around one table, but not at your friend's house or in a restaurant? When is the right time to contradict your boss in the workplace? What are the rules, who decides them and where do I get the manual?

Coming from a complete mix of cultures – Chinese and Welsh – I also know well how exacting and challenging etiquette can be to the unwary observer. If you came across the Pangs eating their 'Sik Fan' Sunday meal – the equivalent of a roast – you might be taken aback by the table manners as we hold the bowls to our faces, shovel the rice and meats into our mouths and slurp the oh-so-flavoursome soup from the bottom of the bowl. Elbows on the table are encouraged, bones are freely spat out onto a designated side plate, and the ritual ends with a massive belch to signify approval of the feast. Yet dare to hold your chopsticks separately in each hand, like a knife and fork, and you will get a clip around the head for your rudeness. Try making sense of that.

I've been reassured that I'm not alone in this. Everyone, apparently, gets nervous about how to behave in particular situations, especially unfamiliar ones. We all have stories about when we've said something we instantly regretted and then dwelled on for weeks. These are the moments you have flashbacks about, even nightmares: being the only one in the room without any clothes on; someone taking offence at an attempted joke. Being embarrassed in public is one of the great human fears. I didn't always know this, but I do now.

Saying or doing the wrong thing is something that both neurotypical and neurodiverse people are evidently capable of, but we probably get there via slightly different means. For someone like me, it's usually through a lack of understanding about social norms and failing to take into account the invisible parameters of hierarchy and convention. If you're neurotypical, you might have suffered the opposite problem: assuming that your knowledge of a certain situation was enough to 'get it right', or overreaching because you felt too comfortable and made a bad joke, or inappropriate suggestion.

Whether your problem was knowing too much or too little etiquette, the solution to social anxiety and faux pas is the same. We need better tools to gather evidence about how people behave and then process that evidence via exploring the various potential outcomes of different behaviour – be that how we dress, what we say, even what kind of greeting we offer. Consider this chapter as a quick guide to avoiding social purdah, from someone who knows exactly how that feels. If you ever worried about how to approach a job interview, to meet your new partner's friends, or before going on a date, then read on.

If the rules are (mostly) unwritten, and no one can agree who sets them, then what can we do to avoid the nightmare scenario of a major etiquette breach? Being someone who is rather fond of a rulebook, I decided that the only way was to write my own. If no one would tell me what the laws of etiquette were, I would have to work them out for myself.

In doing so, relying on techniques from computer modelling, game theory and my own field of bioinformatics, I have learned that a rulebook is perhaps the wrong way to think about etiquette. Because the rules are one thing, and they do exist, but they are not the only variable. It's also about how they are tweaked, interpreted and applied into discrete situations. Individual behaviour is as important as collective habits, and the two influence each other in an unfolding symbiosis that you can never fully predict. To make sense of it, we need techniques to model how local behaviour is both derived from, and different to, the etiquette of a particular country, community or culture. We have to equip ourselves with prior knowledge without letting it restrict us, and explore widely without stepping on people's toes. As I'll explain, when navigating the endless variations of etiquette, the only safe bets are to expect anything, assume nothing and observe everything.

Agent-based modelling

Etiquette is just as challenging to research in theory as it is to navigate in real life. Because it is context dependent, and exists based on varying interpretations, there is no such thing as a universal rule for how to queue, hold a knife and fork, or split a restaurant bill. Before even getting close to the right answer, you need to consider both local norms and individual preferences, which might not even agree with each other.

In other words, etiquette is something that is collectively determined but individually (and selectively) translated. The norms that people sign up to on a national or cultural level then get refracted through an individual, family or workplace prism. To have a chance, we need to tune in at both levels: the shared and the specific. We need a system that can model etiquette both as it exists in theory and in how it is actually applied, inconsistently, by individuals.

Enter agent-based modelling (ABM). This is an approach to modelling that maps complex systems, measuring how 'agents' (which can be people, animals or any other independent actor in a system) behave through their interactions both with the overall system, and with the other agents surrounding them. If you want to know how people are likely to behave in a certain environment – for example, the interactions between traffic and pedestrians, the flow of a Mexican wave at a football game, how consumers will make their way around a shop – ABM is your friend. It's a great way of understanding how actual behaviour – of the agents – relates to expected behaviour – the rules of the system. Ultimately, the system emerges from a balance of the intrinsic rules and the autonomy of its agents, as the two interact with each other.

All of which makes it a great tool for understanding etiquette. ABM reflects the reality that we are both individuals with autonomy, and humans who are subject to numerous constraints on our thinking and behaviour (aka etiquette). We are neither wholly independent, nor total creatures of the system – as ABM shows, the agents respond as much to the behaviour of other agents as to the overall environment. From an analytical perspective, that means understanding the rules of etiquette – how to talk to people, eat a meal and pay a compliment – isn't enough. We also need to observe how people actually interact with those rules and with each other, influencing both in turn. Our behaviour as humans is rarely learned from books (unless you're me, or reading this book) and much more often copied from other people – especially those closest to us. It's how we learned to talk as babies, watching and listening before we could form words ourselves. And it's how we learned the 'right' (in our minds) way to do all sorts of things in life, which may look very strange to others with their own 'correct' way of folding clothes, helping someone in need or cooking a sauce.

This balance between individual and collective, local and global, is the essence of human behaviour. We might think we don't follow the rules but, consciously or not, we are all obeying certain social norms, whether they are universal or specific. Even anarchists have a uniform. But at the same time, frustratingly for the scientist, human behaviour can't be modelled by understanding the rules alone. We also need to scrutinize *how* agents (people) respond to them, and in parallel how the behaviour of other agents influences them.

I use ABM to help me navigate social and professional situations in which I would otherwise be rudderless. It gives me three categories to look at: i) the rules as people say they are, which you can always research in advance, ii) the rules as they are applied in a particular situation, based on how different agents are interacting, and iii) the characteristics and implied preferences of individual agents. ABM helps me to understand that I can't really understand the etiquette of a situation until I have started to experience and observe it. No amount of reading about table manners in Germany, or business culture in Colombia, can prepare me for the realities of the specific context(s) I am going to encounter. Science tells me I will get nowhere until I bridge from learned theory into experimental practice.

You can go into unfamiliar situations equipped with certain knowledge, but it's dangerous to make too many assumptions before you have started to gather actual evidence about your agents – how they interact with both themselves and the system as a whole. I need to observe the agents in action, individually and collectively, to become confident in understanding local etiquette, be that in a particular workplace, household or town centre.

For example, one of the things I find hardest at work is understanding how hierarchy works, and what I should and shouldn't say to different people. (I wasn't kidding about treating everyone the same.) At university, I got 'forcefully resigned' from a part-time job working in IT support, when I publicly contradicted my manager, who was trying to help a customer with a solution I had just tested and knew wouldn't work. After he filed a complaint, I was summoned to see the real boss, who was sympathetic but told me I had to show more respect to people. 'Yeah, but you have to earn it with me', was my retort: exactly how I felt but, as I discovered, not especially conducive to job security.

Now I use a form of ABM to understand how the 'rules' of hierarchy (which in any case are disputed, because lots of companies say they don't really have one, even though most do) actually apply in a workplace. I will take my cue from how the existing agents interact, using that data to work out how to get my own opinions and priorities across. In some environments, people might like the outspoken version of me who says exactly what she thinks, regardless of context or audience. In others, if you want someone to adopt your idea, you're better off presenting it in a way that allows them to think it was theirs anyway. Every system of this kind has its own rhythms and conventions, the etiquette determined by both the preferences and interactions of its agents. You have to first study and then understand these if you are going to find your way through it. It's called *agent*-based modelling for a reason: when you are operating under local conditions, as we almost always are, you have to follow local etiquette. To understand that, you have to focus on the agents, those who do the most to shape behaviour within their sphere of influence. It's what we're always told: the thing that matters isn't so much what you say, but how your intended recipient interprets the message. There is no absolute rule except that of unpredictable individuality.

In using ABM, I have the advantage of being a person who is incapable of making assumptions about anyone or anything I encounter in my life. This can be dangerous, because it means if someone stops me in the street at night, I won't immediately assume that they might mean me any harm. I'll wait to hear what they have to say, and judge their intentions from how they say it. Because in theory I know that this isn't a safe way to behave, I just try to avoid putting myself in this situation – walking alone at night – whenever possible.

As long as you take those precautions, there is a huge benefit to going into a situation without prior assumptions. Unless this is your approach, then you quickly become the victim of confirmation bias, selectively filtering the evidence to fit your preconceived conclusion. Or to put it another way, if you decide in advance that someone is an idiot, then you are going to find reasons to support that conclusion.

The fewer assumptions you can (safely) enter a new situation with, the freer you are to detect the etiquette, and adjust your own behaviour as a result. Try to focus on how the agents in your workplace, networking event or partner's friendship group actually behave, not how you expect them to. Examine them as individuals and track their interactions with the others around them. It's between these individual needs, local connections and global norms that the actual etiquette of a system is to be found.

Game theory

ABM can help you to discover what the etiquette is in a particular context. But it doesn't tell you anything about why people behave this way, or their intentions. Nor does it answer the most pressing question surrounding etiquette: how is someone going to react to what we say or do next? For that we need to delve into the science of game theory, which maps not just how different agents in a system interact, but what their motivations are, and why they make certain decisions.

Game theory was pioneered by two mathematicians whose work helped lay the foundations for the modern study of artificial intelligence: John von Neumann and John Nash. Like agent-based models, it looks at how different players within a certain, rules-based system interact. But it goes further by looking at the consequences of their various choices: how will a decision by one or several players in the game affect everyone else? Game theory looks at the whole picture, assuming a player doesn't just consider their own decisions and their consequences, but those of the other players as well – predicting both what they may know, and how they are likely to act.

Among game theory's many ideas and applications is the Nash equilibrium. This is the concept that, in any finite game, there is a point of balance where all players can take the decision that is in their individual best interests – and that none of them would change that course if the other players' tactics were revealed to them. In other words, equilibrium is reached where individual and collective interests converge and there is no further optimization to be achieved. A proper compromise.

A solution everyone is happy with: be it a playlist, a holiday destination or the food for a picnic.

The Nash equilibrium and its offshoots are used in a wide range of fields, both to understand how aligned or opposed players will approach a particular problem, and to shape policy or decisions which seek to influence the choices of certain players. It's a convergence I have always sought to find between myself and other people – though I'm also fascinated when I can't, and to work out why that is. What's more, whenever a particular group of people changes – be that in its membership or the preferences of the same individuals – then the nature of the Nash equilibrium will evolve in turn.

How does this help us as we tiptoe across the burning coals of social etiquette? Well, to start with it encourages us to look beyond our own perception of certain events, and to put ourselves in the shoes of the other player. Because game theory is ultimately about interdependence – how our outcome depends in part on someone else's choices – we can't just live in our own heads, or base decisions on our own judgement. We need to anticipate how the other person will respond to our question, ice-breaking remark or suggestion. Is what we are about to say or do going to give them the scope to be offended or upset? Based on what we know about the player in question, the context of the interaction and our own ability to execute, how likely is our next move to achieve the desired outcome? What is the effective Nash equilibrium of the situation, in which everyone gets what they want without having to change course?

If ABM allows you to understand the implied etiquette of a given system, game theory is the technique to model your subsequent decisions, aligning them with both your own ideal outcomes, and the choices others are making in parallel or response. It recognizes that you are not just an agent in the system, but a player who must consciously make decisions based on a combination of insight into and ignorance of the rest of the board. We can only navigate by mapping the paths our decisions open up – for ourselves and the other players – and choosing the direction either of Nash equilibrium (mutual benefit) or, if you want to be non-cooperative, individual advancement.

I've come to rely on game theory for explaining *why* certain behaviour exists and overcoming my inability to detect a person's motives (especially as these are rarely made explicit). It might only be hours or days later, discussing it with a friend or family member, that I realize someone said something cruel to me.

Because I'm so bad at instinctively 'feeling' unfamiliar situations, I have to game out every conversation and comment in my head as I go. Often this is a life-saver, moderating the genius comment or helpful observation that would have earned me the eye-rolling response I knew well growing up. But sometimes the algorithm malfunctions and a friendly joke about gingers to my flame-haired Uber driver – carefully calculated to be matey and approachable – leads to instant offence. And that was after making a conscious effort to be friendly, when my preferred, headphones-on approach proved incompatible with a good rating. Another unfortunate mishap was when I wanted to comfort a colleague at work who seemed downcast. Having searched online for appropriate words of encouragement, my chosen phrase – 'Having a good hair day, I see!' – fell flat. I had failed to take into account that he was bald, shiningly so.

For me, game theory is less about winning and more about surviving the life experiences that nothing has prepared me for. I don't want to beat the other players, just to cross the board without sending too many of them flying, like my mum's poor friend at the Christmas party.

This is the counter-intuitive benefit of game theory. While ostensibly being a playbook for rational decision making, it also reminds us of its limits. If we put everything in our lives through the lens of game theory, then we would end up in something like the dystopia that Thomas Hobbes outlined in *Leviathan*, as the fate of humanity without a body politic to bind it together. In its absence, he famously wrote, human life would be 'solitary, poore, nasty, brutish and short': this was the 'state of nature' he believed could only be counteracted by the creation of the centralized state. As game-theory junkies, we would become pure *Homo economicus*: totally self-interested players animated only by our search for what Hobbes characterized as felicity – the pursuit of never-to-be-satisfied desires for power and self-advancement (not, as I thought when I first read it, anything to do with the Austin Powers character Felicity Shagwell).

Game theory could easily become the mechanism to fulfil Hobbes's negative assessment of humans – as creatures that have to be prevented from harming both themselves and others while clambering over each other in a futile attempt to 'win' the game. But it also reminds us of the contrasting opportunity to be *Homo reciprocans*, a person who wants to cooperate with others in pursuit of mutual benefit. The existence of Nash equilibria shows that the ultimate lesson of game theory is interdependence: we are all on the same board, playing the same game, and often depend on the help and support of others to achieve our desired outcome. Game theory could be a selfishness charter, but it's also one of the best frameworks I know for demonstrating how we are all part of the same species, living on the planet, for all our differences essentially sharing the same needs and ambitions.

We don't just learn etiquette for the sake of avoiding social embarrassment. It's also about how we relate to other people and cultures, establishing connection and forging reciprocity. It's the small things that make a difference: picking up rubbish on the street even though it isn't your job; making extra way for a wheelchair user on the street, even though you aren't their carer. It's these minor gestures – the ones that don't immediately benefit us – that make us a social species and not an individualistic one.

Game theory, which doesn't have to be about competition, is one of the most important techniques for finding the common ground that defines our relationships as human beings. After all, if Hobbes's logic stands, then what else has held humanity together up to this point other than the need for cooperation? That might sound warm and fuzzy, but it's also – as cancer reminds us – deadly efficient. Working together isn't just about playing nice, it's also about the most efficient route to goal. And that is why etiquette really matters.

Homology

If agent-based modelling can help us to understand local context, and game theory to plot our own paths alongside those of others, the third leg to my etiquette stool is homology: the science of modelling connections and similarities between disparate data.

While studying other people as agents, and stress-testing decisions in game scenarios, can take you a long way, it doesn't answer all the etiquette questions. What about the things you like to do, and how they will fit into a particular context? How can we be ourselves while staying within the bounds of a particular situation? Why, for instance, is it perfectly OK for me to make a cup of tea sitting on the floor at home (after all, it's the safest place with the least distance for anything to fall), but frowned upon in the office? Or the fact that I like to stir my tea far more times than socially appropriate, since my SPD (sensory processing disorder) relishes the sensation of metal tapping on crockery? What makes it OK for my sister to mock my Frida Kahlo-esque unibrow, but not (I can promise you) for me to point out that her painted-on brows are reminiscent of Super Mario? We need a method for matching behaviour to context and filling in the gaps between our knowledge and ignorance of new situations.

That is where homology, which we use to model the similarities between proteins, comes into its own. Homology is a core technique of bioinformatics, my field of study, where it is used to fill in the gaps in data sets we are still exploring, inferring from related cases. There will always be some missing data, but we can overcome this by using what we know about equivalent situations to inform what we don't about this one. For instance, if you are trying to develop a new drug treatment for a particular form of cancer, and you have found a suitable protein to target, what you need to establish is its structure – the thing you will bind your treatment on to. It's the structure that holds the key, but we probably don't have all the information available. Instead we have to work on parallel information – looking at how our drug binds to other proteins and establishing the areas of similarity. Bit by bit, you can work your way towards a solution – widening the area of known overlap until you have created enough connections to fine-tune your plan of attack.

All we can do is work with the information we have about the current relatives of the target, and existing models which will have areas of overlap. Homology is about bridging from what we do know to make reasonable assumptions about what we don't – creating a map of convergences between known factors: the places our intervention might make the most difference.

In the same way as I use homology at work to better understand the proteins and cells I study, it is my preferred method for trying to establish connections between the scraps of evidence I collect about the people in my life. Drug discovery and the exploration of new human environments share the same basic principle: the evidence is always incomplete, and our ability to reach the right result is dependent on how we navigate from what we do know into the places we don't.

Say I have been dating someone and they want to introduce me to their family. That means I'm entering a new situation, but one about which I have already gathered certain evidence. I will have picked up scraps of information about what their parents and siblings are like; ways in which they are both similar to and different from my boyfriend. Using that I can make certain inferences – their sense of humour, what they might be interested in talking about, things it would be unwise to say or do in their presence. And when the day comes, I will start with the person or topic I feel most confident about: the area of convergence between all the data points that provides me with the safest ground to tread on. Once actually in that environment, I can start to study the agents and model it accordingly, bringing game theory into play to decide exactly what to say and how to act. But it is homology that allows me to take the crucial first steps into the unknown, turning stray bits of data into a hypothesis about new people and situations that makes it safe to enter. It also explains why I generally don't talk to complete strangers: people for whom I have absolutely no evidence base to work with.

In biology, you can never have enough data. It's a bottomless pit because, the more you collect, the more new questions you are having to deal with. Coping with etiquette is no different: your information is never as good as you want it to be, but there is always enough to get started. Homology is about accepting the limits of what you know, and squeezing the most value from the evidence you do have. It's also very revealing of difference and individuality. While my efforts to navigate etiquette began by a desire to know the rules, over time I've realized more and more that it is individual interpretation and nuance that matters most. Just because two friends have blue eyes doesn't mean they will both like carrots. Even within common cultural and social frameworks, it is the differences that make us. Homology allows us to discover what they are, and to better understand both the collective mores and individual quirks that make us the individuals we all are.

If I have learned one thing in my etiquette odyssey, through all my mis-steps, ill-judged comments and collisions with authority, it's that we are all going to get it wrong some of the time. With the best intentions – and modelling techniques – in the world, there is no foolproof recipe for always saying the right thing and avoiding some kind of embarrassment. (Nor should we necessarily want to – think of all the stories we'd never get to tell.)

My advice is to forget about perfect when it comes to new social and professional situations. Instead, focus on lowering your error count – and counting the little things you have achieved (which for me is only annoying a maximum of two people in a twenty-four-hour period).

Use the techniques of observation, calculation and connection I have outlined here to feel your way into new situations, stepping only onto ground where you have some level of confidence. Don't obsess about your errors (easier said than done, I know) and instead focus on what you have learned. There will always be something you don't know; because the more you learn, the more you have to discover. The game of etiquette is endless: you will never complete it. But it's not a competitive game – in fact it's more about delaying your immediate needs for someone else's sake, and that of mutual benefit.

Above all, remember it's not necessarily about what you say or do, but the impression you leave behind on people, and how you would like to be remembered. Even if you get it wrong, the effort of having tried is itself worth something. People receive the signal, even if they don't appreciate the gesture itself. Better to turn up with a watermelon someone doesn't want than with nothing at all.

Afterword

As I've worked through this book, reflecting on all the experiences that have brought me to this point, I've been trying to identify when it all changed. I know that there was a moment, around the age of seventeen, when I started to feel human for the first time. Not all the time, and often only for a fraction of a second. But as someone who had always felt like an outcast, it was transformative. Suddenly there seemed to be more colour, the haze in my head would lift and the world of confusion surrounding me would briefly start to make sense. All the experiments I had trialled, all the quasi-algorithms I had developed for myself – suddenly they started to work. The pieces began to fit together. I was in phase.

But when I think back on this, I can't remember when the first of these moments was; and I don't know exactly what triggered it. Like a plant flowering in the springtime, my sense of being human was something I could only see and enjoy after the fact. I knew it when I felt it. I just hadn't realized that I was getting there, or at what speed.

I'm still not 'there', and I think I probably never will be. Part of me will always be on my own island, and I'm happy with that. (If you own an island, why would you sell, right?) But what I've learned is that it's possible to change yourself: not to deny or erase your true self, but to improve – getting better at the complex business of being human, in the way we plan our lives, manage our days, balance our emotions and nurture our relationships.

I've also learned (I think) what it takes to do this. And, in a word, that's patience.

This is perhaps the biggest of my many contradictions. My ADHD brain is one of the least patient beings in existence. But as a person, and especially as a scientist, I can be painstakingly patient. I know from experience that good things don't happen quickly, experiments never succeed the first time, and it's only by failing and using what you've learned that you can make progress.

Of course this hasn't come easily and is something I still struggle with. It's taken a lot of hysteria, emotional outbursts and procrastination to get to the point where I can not only see the value of patience but also sometimes even embody it. I've worked for this; and it's been worth it.

The greatest correlation of all between science and living is that they are both in equal parts frustrating and rewarding for those who persevere. There is nothing in my life that thrills me like a breakthrough in the lab: that moment when a door finally opens towards the solution you have been seeking. It's the novelty of discovery, however small, that means I love my work so much. Any scientist will tell you the same.

As this book has detailed, I have taken this same approach into working out how to live and function better as a human. And I think everyone can benefit from a little of the same. There are things we would all like to improve in our lives – to feel more human connection, to sharpen our ambitions or improve the way we pursue them.

This is possible, but it's not easy. The mind and body are like an athlete that needs to be trained to improve perception, memory, processing and empathy. It's a progression you can't expect or demand to yield quick results, any more than you would in the gym. These are some of the most fundamental things about us as people, and you won't transform them overnight. But if you want to, and you're willing to show an athlete's commitment, it's all eminently possible. The concepts and techniques I've outlined are essentially disciplines: they can be useful, but only if they are trained and embraced continuously over time. It's attritional, just like science. Like everyone else, I am the product of my failed experiments: proudly so.

Growing as a person is incredibly frustrating, because we put in all this work, and for a time – maybe a long time – nothing happens. At this point it's easy to lose heart and give up. But the real reward lies in persevering, pushing past the uncertainty and self-doubt until one day change has crept up on you. We don't get to plan how or when this happens. We can only put in the work, and trust in the process.

So don't despair at the next unrealized plan, unfulfilled goal or failed relationship. Learn from them. Try something different the next time. Experiment into how to do things your way. Accept the very human inevitability that getting better at living is a slow and gradual process. And, whatever happens, don't demonize the things that make you different. Embrace them, like I have, as your innate superpowers.

It's going to go wrong before it goes right. To get worse before it gets better. That's OK – in fact, it's essential. Relish your failed experiments. Have fun working it out on your own. And don't apologize for being yourself. I never have, and I don't intend to start now.

Acknowledgements

Forever thankful for the book team – the ones who saw me.

These are the people who enabled this idea of mine, and my series of notebooks, to come alive: Adam Gauntlett, Josh Davis, Emily Robertson.

My teachers and mentors – the ones who supported me at school and beyond.

For their relentless patience in explaining subjects to me, inspiring and believing in me. Regardless. My teachers: Keith Rose, Lorraine Paine and Margie Burnet Ward. My mentors: Michelle Middleton, Allyson Banyard, Clare Welham, Lesley Morris, Celia Collins, Katy Jepson, Leo Brady, and my PhD supervisor Christine Orengo.

I am eternally grateful for my friends – the ones who have seen it all.

For Abigail – my extra sister, one of my best buddies, and the person who gave me the confidence to put my book out there. For the people in the lab (also known as my protein family) who backed me. For Maísa, Elodie, Bruna, Amandine, Pip, Sam and Tina, who provided constant support and encouragement, and one of my oldest friends, Rosie. Greg – for telling me that a bad turn of events always makes for a good story, and Rhys – who told me to never stop writing.

My family – the ones who have *really* seen it all.

For Sonia, Peter, Lydia, Roo, Nay, Rob, Jim, Tiger, Lilly, Aggie, and the Pang family. My cousins Lola, Ruby, Tilly, Aunty Sue and Tina, Uncle Rob and Huw, and a special note to Uncle Mike and Uncle John for turning a blind eye when I permanently borrowed their science books, that in turn fed this whole process. Much gratitude also goes to the Pang grandparents – Cheung Fook, Sui Ying – and in loving memory of the Anslow grandparents – Francis and Elizabeth (Betty). These people are home to me. I am always reminded where I have come from, and to carry on embracing what makes me different and do what makes me tick. Without their support, I am not actually sure whether I would be here or not. Thank you all for everything.

Chapters dedicated to:
  1) The book crew, Josh Davis and Emily Robertson (editors) and Adam Gauntlett (agent)
  2) Fellow scientists
  3) Mother Sonia
  4) Mentors
  5) Father Peter
  6) Fellow hipsters
  7) Sister Lydia
  8) Fellow Aspies
  9) Friends who have been, friends who are, and friends who will be
  10) Little me
  11) Strangers who chucked me a bone

Index

Page references in *italics* indicate images.